Work in the Fut

Robert Skidelsky • Nan Craig
Editors

Work in the Future

The Automation Revolution

Editors
Robert Skidelsky
Centre for Global Studies
London, UK

Nan Craig
Centre for Global Studies
London, UK

ISBN 978-3-030-21133-2   ISBN 978-3-030-21134-9   (eBook)
https://doi.org/10.1007/978-3-030-21134-9

© The Editor(s) (if applicable) and The Author(s), under exclusive licence to Springer Nature Switzerland AG 2020
This work is subject to copyright. All rights are solely and exclusively licensed by the Publisher, whether the whole or part of the material is concerned, specifically the rights of translation, reprinting, reuse of illustrations, recitation, broadcasting, reproduction on microfilms or in any other physical way, and transmission or information storage and retrieval, electronic adaptation, computer software, or by similar or dissimilar methodology now known or hereafter developed.
The use of general descriptive names, registered names, trademarks, service marks, etc. in this publication does not imply, even in the absence of a specific statement, that such names are exempt from the relevant protective laws and regulations and therefore free for general use.
The publisher, the authors and the editors are safe to assume that the advice and information in this book are believed to be true and accurate at the date of publication. Neither the publisher nor the authors or the editors give a warranty, expressed or implied, with respect to the material contained herein or for any errors or omissions that may have been made. The publisher remains neutral with regard to jurisdictional claims in published maps and institutional affiliations.

This Palgrave Macmillan imprint is published by the registered company Springer Nature Switzerland AG.
The registered company address is: Gewerbestrasse 11, 6330 Cham, Switzerland

Contents

1 **Introduction** 1
Robert Skidelsky and Nan Craig

2 **The Future of Work** 9
Robert Skidelsky

Part I Work in the Past 23

3 **Patterns and Types of Work in the Past: Part 1** 25
Richard Donkin

4 **Patterns and Types of Work in the Past: Part 2** 33
Richard Sennett

5 **Patterns and Types of Work in the Past: Wageworker and Housewife from a Global Perspective: Birth, Variations and Limits of the Modern Couple** 37
Andrea Komlosy

Contents

Part II	**Attitudes to Work**	51
6	**Attitudes to Work and the Future of Work: The View from Economics** *David A. Spencer*	53
7	**Attitudes to Work** *Pierre-Michel Menger*	65
8	**Work as an Obligation** *Nan Craig*	73
Part III	**Attitudes to Technology**	81
9	**Attitudes to Technology: Part 1** *James Bessen*	83
10	**Attitudes to Technology: Part 2** *Carl Benedikt Frey*	89
Part IV	**Possibilities and Limitations for AI: What Can't Machines Do?**	97
11	**What Computers Will *Never* Be Able To Do** *Thomas Tozer*	99
12	**Possibilities and Limitations for AI: What Can't Machines Do?** *Simon Colton*	109

Part V Work in the Digital Economy   123

13 Work in the Digital Economy   125
Daniel Susskind

14 Two Myths About the Future of the Economy   133
Nick Srnicek

Part VI AI, Work and Ethics   143

15 AI, Ethics, and the Law   145
Cathy O'Neil

Part VII Policy   155

16 Policy for the Future of Work   157
David Graeber

17 Automation and Working Time in the UK   175
Rachel Kay

18 Shaping the Work of the Future: Policy Implications 189
Irmgard Nübler

Index   203

Notes on Contributors

James Bessen is an economist who studies technology and innovation policy. He has also been a successful innovator and CEO of a software company. Bessen is Executive Director of the Technology and Policy Research Initiative at the Boston University School of Law. Bessen's book *Learning by Doing: The Real Connection Between Innovation, Wages, and Wealth* looks at history to understand how new technologies affect wages and skills today. Prior research on innovation includes the book *Patent Failure*, written with Michael Meurer. Bessen's research has been widely cited in the press as well as by the White House, the US Supreme Court and the Federal Trade Commission.

Simon Colton is Professor of Computational Creativity at the School of Electronic Engineering and Computer Science, Queen Mary University of London, and SensiLab, Faculty of Information Technology, Monash University, Australia. Previously, he was a Reader in Computational Creativity in the Department of Computing at Imperial College, London.

Colton is an Artificial Intelligence researcher, specialising in questions of Computational Creativity. He leads the Computational Creativity

Group, and is also behind The Painting Fool, a computer program that he hopes will one day be taken seriously as a creative artist in its own right.

Nan Craig is Programme Director at the Centre for Global Studies and researches technology and the future of work. She holds an MSc in Global Politics from the LSE and previously worked for the social enterprise Participle, and as a freelance researcher. She also writes fiction which has been published by the *New Scientist*, *Vice* and *Magma*.

Richard Donkin is an author and commentator on work and management. A former columnist and writer at the *Financial Times*, he is the author of two books, *The History of Work* and *The Future of Work*, both published by Palgrave Macmillan. He is a visiting fellow at Cass Business School.

Carl Benedikt Frey is Oxford Martin Citi Fellow at University of Oxford where he directs the programme on the Future of Work at the Oxford Martin School. He is author of *The Technology Trap: Capital, Labor and Power in the Age of Automation.*

Frey has served as an advisor and consultant to international organisations, think tanks, government and business, including the G20, the OECD, the European Commission, the United Nations and several Fortune 500 companies. He is also an op-ed contributor to the *Financial Times, Scientific American* and the *Wall Street Journal.* In 2016, he was named the second most influential young opinion leader by the Swedish business magazine *Veckans Affärer.*

David Graeber is an American anthropologist, activist and author of *Debt: The First 5000 Years* (2011) and *Bullshit Jobs: A Theory* (2018). He is Professor of Anthropology at the London School of Economics.

Rachel Kay is a researcher at the Centre for Global Studies. Before joining the Centre, she completed an MPhil in Development Studies from the University of Cambridge. During her MPhil she researched the trade of bazaar goods between China and the Central Asian states, focusing on the transnational relationships created by bazaar trade.

Andrea Komlosy is a professor at the Institute for Economic and Social History, University of Vienna, Austria, where she is coordinating the Global History and Global Studies programmes. She has published on labour, migration, borders and uneven development on a regional, European and global scale, recently "Work and Labour Relations", in Kocka Jürgen/van der Linden Marcel (eds.), *Capitalism: The Re-Emergence of a Historical Concept* (2016) and *Work: The Last 1000 Years* (2018).

Pierre-Michel Menger studied philosophy and sociology at the Ecole Normale Supérieure in Paris and obtained his PhD at the Ecole des Hautes Etudes en Sciences Sociales in 1980. He was a senior researcher at the Centre National de la Recherche Scientifique (Paris) before joining the Collège de France, where he is a professor since mid-2013, holding the Chair of Sociology of Creative Work. He is also professor (directeur d'études) at the École des Hautes Etudes en Sciences Sociales. He is the author or co-author of 15 books, and his numerous published articles have appeared in journals such as *Revue française de Sociologie*, *Sociologie du travail*, *L'Année Sociologique*, *Annales*, *Annual Review of Sociology*, and *Poetics*. He was editor of the *Revue française de Sociologie* and is a member of the editorial boards of the *Revue Economique*, *Revue française de gestion* and *Genesis*, among others. His book *The Economics of Creativity* was published in 2014.

Irmgard Nübler is a senior economist in the Research Department of the International Labour Organisation (ILO) in Geneva. She leads the research programme on technology productive transformation and jobs. In previous capacities in the ILO she was responsible for developing work programmes on skills development and apprenticeship systems, and on industrial policies and capabilities for job creation. Before joining the ILO, she was a professor of Economics at the Free University of Berlin and conducted research, among others, at the Massachusetts Institute of Technology in Boston and the Institute for Development Studies at the University of Nairobi. She holds a PhD in Economics from the Free University of Berlin.

Cathy O'Neil received a PhD in Math from Harvard, was a postdoc at the MIT math department, and a professor at Barnard College where she published a number of research papers in arithmetic algebraic geometry.

She then switched over to the private sector, working as a quant for the hedge fund D.E. Shaw in the middle of the credit crisis, and then for RiskMetrics, a risk software company that assesses risk for the holdings of hedge funds and banks. She left finance in 2011 and started working as a data scientist in the New York start-up scene, building models that predicted people's purchases and clicks. She wrote *Doing Data Science* in 2013 and launched the Lede Program in Data Journalism at Columbia in 2014. She is a regular contributor to Bloomberg View and wrote the book *Weapons of Math Destruction: How Big Data Increases Inequality and Threatens Democracy*. She recently founded ORCAA, an algorithmic auditing company.

Richard Sennett trained at the University of Chicago and at Harvard University, receiving his PhD in 1969. Over the course of the last five decades, he has written about social life in cities, changes in labour and social theory. Among other awards, he has received the Hegel Prize, the Spinoza Prize, an honorary doctorate from the University of Cambridge and the Centennial Medal from Harvard University. He teaches sociology at New York University and at the London School of Economics.

Robert Skidelsky is Emeritus Professor of Political Economy at Warwick University. His three-volume biography of John Maynard Keynes (1983, 1992, 2000) won five prizes and his book on the financial crisis—*Keynes: The Return of the Master*—was published in September 2010. He was made a member of the House of Lords in 1991 (he sits on the crossbenches) and elected a fellow of the British Academy in 1994. He is also the co-author of *How Much is Enough? The Love of Money and the Case for the Good Life* (2012), written with his son Edward, author of *Britain in the 20th Century: A Success?* (2014), and editor of *The Essential Keynes* (2015). He has recently filmed a series of lectures on the history and philosophy of economics which will be made available as an open online course in partnership with the Institute for New Economic Thinking. His most recent book, *Money and Government: A Challenge to Mainstream Economics*, was published in 2018.

David A. Spencer is Professor of Economics and Political Economy and Head of the Economics Division at Leeds University Business School, the University of Leeds. His research focuses on the study of work, especially within economics. He retains broader research interests in political economy and the history of economic thought.

Nick Srnicek is a lecturer in Digital Economy at King's College London. He is the author of *Platform Capitalism* (2016) and co-author of *Inventing the Future* (2015 with Alex Williams). With Helen Hester, he is finishing his next book *After Work: The Fight for Free Time* (2020).

Daniel Susskind explores the impact of technology, particularly artificial intelligence, on work and society. He is a fellow in Economics at Balliol College, University of Oxford, where he teaches and researches. He is the co-author of the best-selling book *The Future of the Professions*. His TED Talk, on the future of work, has been viewed more than 1.4 million times.

Previously Susskind worked in the British Government—as a policy adviser in the Prime Minister's Strategy Unit, as a policy analyst in the Policy Unit at 10 Downing Street, and as a senior policy adviser in the Cabinet Office. He was a Kennedy Scholar at Harvard University.

Thomas Tozer graduated from the London School of Economics with an MSc in Philosophy and Public Policy in 2015. Since then, he has worked as a researcher at the Centre for Global Studies and the Intergenerational Foundation, and as a policy advisor in the Civil Service. Outside of work, he has appeared on Sky News and spoken on BBC Radio 5 Live to discuss the impact of government policy on young people. His published essays include "Increasing Electoral Turnout Among the Young", "A New Intergenerational Contract" and "Brexit, Democracy and Intergenerational Justice". In his free time, Tozer enjoys cycling, drinking tea and practising Kadampa Buddhism.

1

Introduction

Robert Skidelsky and Nan Craig

When we planned a symposium in February 2018 on the future of work, we divided the subject into eight areas. We hoped to cover more ground than is usual, and to look at how work has changed in the past as well as how it is changing now and in the future. Most of the contributions to this book came out of that symposium, and reflect their original beginnings as oral presentations. Other pieces were commissioned later in order to extend the thematic reach of the book even further.

When we talk about the future of work, too often the discussion is narrowly focused on automation, and the social or economic problems that are assumed to arise from it. What this collection of essays aims to do is to broaden that discussion. How has the character of work changed in the past, and what can that tell us about how it will change in the future? How have our attitudes to work shifted over time? How will increasing automation over the coming decades—of professional as well as routine or manual work—change our relationships with work and

R. Skidelsky • N. Craig (✉)
Centre for Global Studies, London, UK
e-mail: skidelskyr@parliament.uk

with each other? Finally, what kind of actions can we take in response to these changes?

The effect of automation on human work has been almost constantly in the headlines during the past few years—perhaps in part simply because slogans about 'robots taking over' make good copy. Whether reality lives up to the headlines is less certain. Technology destroys jobs, but in the past it has created new jobs to replace the ones it has destroyed. This could be the case in the future. In the past, it has not only created new jobs but reduced the hours of work per job. This could also repeat itself. On the other hand, there is the view that, whatever may have been true in the past, we have now reached a tipping point—or soon will—when the advent of intelligent machines is simply going to destroy existing jobs faster than it creates new jobs. If so, technological unemployment would turn from its relatively benign past process into a virulent involuntary one.

There are several issues worth discussing, keeping those two views in mind. First of all is history. What do the long run data of population growth, employment growth, hours of work, earnings per hour worked since the industrial revolution tell us? What do they actually show? Most people treat the Luddite fear of net job loss as a prediction that turned out wrong, but *why* they were wrong and how wrong were they? Given careful ceteris paribus conditions, the Luddites were correct, as indeed David Ricardo recognised in his essay 'On Machinery'.

According to a recent McKinsey Global Institute report, 50% of time spent on human work activities in the global economy could, theoretically, be automated today, though the current trend suggests a maximum of 30% by 2030, depending on speed. However, estimates of jobs at risk tell one nothing about net job outcomes.

The most widely held view is that there will be net job losses, or technological employment, but these will be temporary or transitional. There is the old economists' distinction between the short run and the long run; no one ever specifies how short the 'short' run is. Against that view that the job losses will only be transitional are more pessimistic views from people like Martin Ford and Larry Summers. They suggest that in fact job losses will be permanent unless something is done.

Another issue is where new jobs are to come from. Categories of human jobs widely expected to maintain themselves or expand in line with the contraction of others are creative jobs, jobs requiring exceptional manual dexterity, person to person services, notably healthcare, care work and so on. How many of these jobs will be created? Why should their number equal the total of jobs automated? For creative industries, a winner-takes-all projection is quite common. Top artists get top pay and ordinary ones get nothing, or almost nothing.

The next issue: the question of how much people will want to work, or need to work, depends not only on technology and the nature of future work, but on what we think about human wants and needs. Needs and wants are not identical, though they are treated as such by economists. We usually want what we need, but we by no means need all we want. The question of how much we will want to work in the future partly depends on a view we take about human nature and the drivers of consumption.

We began the symposium by looking into the past—in particular at the patterns of pre-modern work and how work has changed in the past few centuries. Pre-industrial work may have been arduous, but it was much more intermittent than modern work, since activity—certainly agriculture, but even fighting—was seasonal. What did work mean to people? The definition of work has narrowed in the twentieth century to paid employment, setting up a false dichotomy between work and leisure. Have we always distinguished between homo ludens and homo laborans? We have come to think of these as opposites, but it is not always so and may not be so in the future.

Richard Donkin stresses how for much of human history, work was inextricable from the rest of life—not only as a means of survival but also as an element of family or social life and as a pleasurable activity.

Richard Sennett looks to the craftsmanship of the past to show us how important the physical body is to mental labour, and how creativity of intellectual work is as reliant on physical processes as it is on mental effort. This belies the argument that automated systems can effectively replace human labour, and suggests that we should be cautious in adopting what appear to be labour-saving technologies which, by virtue of removing the physical element of intellectual work, undermine and hollow out skills.

Andrea Komlosy's chapter charts the rise and fall of the gendered wage-earner/housewife model of work which took over from household economies during the industrial revolution, and how it spread from Western Europe, but failed to dominate in other areas of the globe. When neither men nor women want to or can afford to be purely 'home-makers' or 'breadwinners', how can work in the home and outside it be shared?

Having looked at some of the ways in which patterns of work have changed over time, we turn to changes in people's attitudes to their work. As Studs Terkel put it in his oral history, *Working*, work is 'about a search, too, for daily meaning as well as daily bread […] in short, for a sort of life rather than a Monday through Friday sort of dying'.

David A. Spencer explains how the attitudes of economists to work have influenced wider cultural narratives around this activity—in particular, the inability of classical and neoclassical economics to conceive of work as anything other than arduous and costly to the worker. Pierre-Michel Menger describes how positive and negative cultural attitudes to work can interact within one society—specifically, in France. Nan Craig traces the historical antecedents of our attitudes to work and suggests that we can broaden the definition of work again, and find ways to accommodate other kinds of work than waged jobs.

Next we directly address the technological developments that are giving rise to changes in working life. Both Carl Benedikt Frey and James Bessen are sceptical about the idea that advances in automation are likely to have either unequivocally positive or negative effects on working lives. Carl Benedikt Frey explains how attitudes to technological development and automation are likely to be affected by how that technology affects people's working lives—and that the adoption of the technology itself will in return be mediated, in part, by those attitudes.

James Bessen describes the ways in which automation has changed jobs, and in what ways this is likely to continue or develop. He argues that specific jobs or categories of work rarely become obsolete in their entirety—rather, they evolve as technology becomes available to automate particular elements of the work. In deciding whether automation creates or destroys jobs, the decisive factor is demand rather than technology.

What are the advantages of technology? Often people say, 'It is obvious that self-driving cars will reduce the numbers of road accidents. Automated diagnostic and treatment systems will reduce medical casualties and so on'. We know that argument, but will algorithmic trading increase the efficiency of financial markets, or render them more liable to crashes? So far, it seems the latter has been the case. We also need to scrutinise the more generalised idea that technology increases human welfare by increasing the affordability and thus availability of consumption goods. This invokes all kinds of questions about the relationship between consumption and happiness, and the damage done to the planet by the constant pursuit of material wealth.

Is technology determinative? Even technological utopians assume technological invasion takes place in a social and economic context, which determines what is invented, how quickly inventions are applied and so on. Historically, inventions did not necessarily become widely used; the history of technology, up until the early modern period, was patchy rather than progressive. Automata, for instance existed in the ancient world, but they were novelties for kings and did not prompt wider technological change.

Implicit in modern arguments is technological determinism. It underpins nearly everything currently being said about automation and we need to tease out its implicit assumptions. Is it true that technology is like a runaway train over which we have no control once it has left the station, and the only thing we can do is to adapt to its demands? This is certainly the implication of the 'robots will take your job' rhetoric.

In response to this, Thomas Tozer takes a theoretical view of whether artificial intelligence can truly replicate the abilities of humans. In contrast, Simon Colton argues that it is a mistake to compare human and machine intelligence, and that machines can be capable of intelligence and creativity without necessarily being conscious or having specifically human attributes.

In the next section, Daniel Susskind and Nick Srnicek address misconceptions about the changes that result from the shift towards an automated, digital economy. Daniel Susskind reflects on how old assumptions have proved unreliable on estimating which tasks can be automated and which cannot. Nick Srnicek argues that there are two key misunderstandings of platform economies. First, the idea that Uber's business model will serve as a model for the rest of the economy is too simplistic. In that sense, both the promise and the threat of 'platform capitalism' to jobs can be overstated. Secondly, he argues that AI does pose a threat to the economy, not because of automation but because it further encourages the tendency towards monopoly that already exists in the platform economy. Again, easy assumptions about the future cannot be relied upon.

Cathy O'Neil was asked to address the ethical dimensions of AI's effect on working life. She gives three examples from the legal field to illustrate the ethical dangers of using algorithmic solutions unthinkingly or without careful oversight. In particular there is a danger that patterns of decisions made by algorithms can become self-reinforcing, as their effect on reality starts to feed back to them, creating a loop.

Finally, what measures might be useful in alleviating problems caused by automation? Of course, the way we react to change depends not only on the problem but also on the assumptions we hold about what would constitute a good outcome. Both David Graeber and Rachel Kay favour reducing human work, while Irmgard Nübler focuses on how technological unemployment can be mitigated and job growth maintained.

David Graeber argues that the future of technological unemployment predicted by J.M. Keynes has in fact come to pass—but that we have compensated for the lack of work by creating millions of make-work jobs with little purpose. He recommends giving people the means to leave pointless jobs by severing livelihood from work through a universal basic income.

Rachel Kay takes a different tack, discussing the argument for reducing working hours. Workers in the UK work longer hours than in other European countries; looking at Germany, France and the Netherlands as examples, she makes recommendations on how the UK could move in the same direction.

Irmgard Nübler analyses the way that different types of innovation and market forces affect job creation and destruction, as well as how those forces can be harnessed to support job creation. She notes that as well as intervention by governments to shape workforce skills and aid adjustment to new technologies, we need redistributive policies to ensure that productivity gains are more evenly shared.

What the wide range of approaches here show is that not only is the future uncertain, but our *aspirations* for the future vary widely, too. It will be complicated enough to deal with the potential economic and ethical pitfalls associated with the growth of Artificial Intelligence (AI) technology, but we also need to know what kind of future we are aiming for.

We hope this book will contribute to that debate.

2

The Future of Work

Robert Skidelsky

A society, wrote Jan Patočka, is decadent if it encourages a decadent life, 'a life addicted to what is inhuman by its very nature'.[1] It is in this spirit that I want to explore the impact of technology on the human condition, and especially on work. Is technology making the human race redundant materially and spiritually—both as producers of wealth and producers of meaning?

For an optimistic answer to this question, let me turn to John Maynard Keynes. In his 1930 essay, 'Economic Possibilities for our Grandchildren', Keynes thought that technological progress would produce so much extra wealth that in about 100 years or even less, we would be able to reduce working hours to just 15 a week, or 3 a day. This process, he warned, was unlikely be smooth.

---

[1] *Heretical Essays in the Philosophy of History*, 97.

R. Skidelsky (✉)
Centre for Global Studies, London, UK
e-mail: skidelskyr@parliament.uk

We are being afflicted with a new disease … namely *technological unemployment*. This means unemployment due to our discovery of means of economising the use of labour outrunning the pace at which we can find new uses for labour. But this is only a temporary phase of maladjustment. All this means in the long run *that mankind is solving its economic problem*.[2]

Ever since machinery became an active part of industrial production, redundancy has been seen either as a promise or a threat. The former has been the dominant discourse in economics, with redundancy seen as a transitional problem, confined to particular groups of workers, like the handloom weavers of early nineteenth century Britain. Over time, part of the displaced workforce would be absorbed in new jobs, part of it in the greater leisure made possible by improved productivity.

However, the fear of the permanent redundancy of a large fraction of the workforce—that is, its forced removal from gainful employment—has never been absent. The reason is that the loss of human jobs to machines is palpable and immediate, whereas the gain is indirect and delayed: an immediate threat versus a long-term promise.

The fear of redundancy has two roots. The first is people's fear that machines will rob them of their livelihood; the second that it will rob them of their purpose in life. Sociologists stress the importance of work in giving meaning to a person's existence. Economists, on the other hand, see work as purely instrumental, a means for buying things people want. If it can be done by machines, so much the better—it may free up people for more valuable pursuits.

It is not surprising that fear of redundancy surfaces whenever there is a burst of technological innovation. We are living through such a period now with the spread of automation. The headlines tell us that robots are gobbling up human jobs at an unprecedented rate—that up to 30 per cent of today's work will be automated within 20 or so years. And the jobs themselves are becoming ever more precarious. So the old question is being posed ever more urgently: are machines a threat or a promise?

---

[2] Quoted from Robert Skidelsky ed. *John Maynard Keynes: The Essential Keynes*, 80. Italics in original.

Work in History

The western concept of work starts with the 'disdain for work' of the ancient world. Working for a living was despised. The good life was one devoted to politics in the Greek conception and to self-cultivation in the Roman. This ideal depended on slaves doing what we call work. Slaves, said Aristotle, were tools, and were tools by nature. Presciently, though, he speculated that one day mechanical tools might replace human tools.

The ancient contempt for work could not survive the decay of slavery, though vestiges of it have lingered in all aristocratic societies. This explains not just the high approbation of leisure by the elite of Keynes's day, all whom were educated in the classics, but also the hostility to the ideal of leisure by the majority who associated with it not just the idleness of the rich, but with unemployment. The Chicago School notion that being out of work represents a 'choice for leisure' is the conceit of economists who have never experienced a day's unemployment in their lives.

With Judeo-Christianity the story gets more complicated. Work is the 'primal curse', the punishment by God for Adam's sin of eating the forbidden fruit of knowledge; but at the same time it is a divine injunction to cultivate the fruits of the earth God has bestowed. Properly understood, God's punishment was not to make people work, but to make work painful. Anthropology reflects the Biblical story in the idea of 'original affluence'. Hunter gatherers needed everything around them, but they had everything they needed. Hunting was not work. Work enters human history with agriculture: 'in the sweat of thy face shalt thou eat bread' (Gen: 3:19). The monastic economy was built on the precept *ora et labora*: pray and work.

The productive unit in the pre-modern economy was the household not the factory: work and life were not yet separated. The division of labour was internal to the household. Unlike the ancients, the Middle Ages attached value to craftsmanship or 'making things'. The medieval economy comprised farms and 'manufactories' in small towns which were little larger than villages. The professions had their origin in the urban guilds of skilled workers. Yet everyone was skilled in the sense that their work involved knowledge of all stages of production, not just tiny bits of it, as in Adam Smith's pin factory. The nature of the work forced them to be 'multi-tasked'. Associated with multi-tasking was multiple sources of income, with the 'putting out' system providing extra income for farmers in the fallow season.

The social and political structure was hierarchical: everyone had their place and their just reward in the scheme of things. Religion offered solace for earthly suffering. Universal insatiability, emphasised by economists as fundamental to human nature, was still in the future. In the late medieval world, commerce spread within Europe and to other lands, but it was still confined to the fringes of the economy. Temporary and permanent redundancy of the population there certainly was—but this was caused by harvest failures, wars, or plagues, not by competition from machines.

With the Enlightenment, the idea of work came to be associated not with the husbanding of nature, but with 'overcoming' it, the human project which has dominated western history ever since. It was human participation in this project, made possible by science, which was supposed to set the whole of humanity free, and not just that small minority of the wealthy and powerful. This was the democratic promise of work.

The particular form of progress which excited the eighteenth century imagination was the growth of wealth. 'The end of production is consumption' wrote Adam Smith. The more goods there were, the happier we would all be. The new human project spurred invention. The accumulation of wealth required machinery, since human work alone could not wrest more than a limited amount of produce from the earth. The Industrial Revolution changed the human link to work in a profound way. It replaced the artisan by the mechanic and home production by factory production. We have entered the age of capitalism, the economists and economic motives.[3]

---

[3] See e.g. Richard Donkin, *The History of Work*; Andrea Komlosy, *Work: The Last Thousand Years*, ch. 1; Keith Thomas, *The Ends of Life: Roads to Fulfilment in Early Modern England*, 91ff.

The Economists and Machines

The fact that work was held to be the necessary means to enjoyment did not mean that it was itself enjoyable. The economists' conception of labour shed any idea that it was natural or intrinsically satisfying. Work was not a curse, but it was a cost—the cost of consumption. For without work there would be no money to buy things. As Lenin was to put it with his customary bluntness: under communism *Kto ne rabotaet, tot ne est*: 'Who does not work shall not eat'.

By the same token though, any reduction in this cost by the use of machinery opened up a brighter future: more output and therefore more money for less effort. The increase and improvement of machinery was inextricably linked both to the denial of satisfaction in work and to the promise of more and better consumption. Economists unanimously welcomed the dawn of the machine age.

As David Ricardo explained in the first edition of his *Principles of Political Economy and Taxation* (1817):

> If, by improved machinery, with the employment of the same quantity of labour, the quantity of stockings could be quadrupled, and the demand for stockings were only doubled, some labourers would necessarily be discharged from the stocking trade; but as the capital which employed them was still in being, and as it was the interest of those who had it to employ it productively, it appeared to me that it would be employed on the production of some other commodity useful to society, for which there could not fail to be a demand; for I was, and am, deeply impressed with the truth of the observation by Adam Smith, that 'the desire for food is limited in every man by the narrow capacity of the human stomach, but the desire of the conveniences and ornaments of building, dress, equipage, and household furniture, seems to me to have no limit' And, then, as it appeared to me that there would be the same demand for labour as before, and that wages would be no lower, I thought that the labouring class would, equally with the other classes, participate in the advantage, from the general cheapness of commodities arising from the use of machinery.[4]

---

[4] *Principles*, 3rd ed., ed. D. Winch, 1973, 274.

Two features of Ricardo's argument have long been part of the economics of innovation: first, the treatment of supply (capital and labour) as malleable or fluid, based on the assumption of wage flexibility, geographic mobility and easy transferability of skills. Workers employed in producing stockings can be squeezed almost costlessly into the right shape for producing widgets. Technically, in other words, labour is treated as homogeneous. Second, the assumption of insatiability as the key motive driving forward the progress of wealth. In fact, Ricardo anticipates Lionel Robbins's famous definition of economics as the science that studies human behaviour as a 'relationship between ends and scarce means which have alternative uses'.[5]

However, by the third edition of his Principles in 1821, Ricardo had somewhat changed his tune. The interval had seen the most intense period of the Luddite disorders. The Luddites, as is well known, were groups of English weavers who destroyed factory machinery—wide knitting frames and power looms—in the early days of the Industrial Revolution. They were eventually put down by the military, and their leaders hung or transported to Australia, despite Lord Byron making an eloquent speech in their defence in the House of Lords.

Scientific economic theory held the Luddite argument to be economically illiterate. But in 1821 Ricardo added a 31st chapter, entitled 'On Machinery', which was heavily influenced by the work of a now unknown economist, John Barton. In his *Observations on the Circumstances which Influence the Conditions of the Labouring Classes of Society* (1817), Barton argued that capitalists could just as well invest their profits in new machines as in additional labour. In doing so, he displayed a clear understanding of the principle of substitution between labour and capital, way ahead of his time.[6]

Ricardo in effect accepted Barton's argument. Two conclusions from his chapter 31 have been debated ever since: first, Ricardo's statement that the opinion prevailing in 'the labouring class, that the employment of machinery is frequently detrimental to their interests, is not founded on prejudice and error, but is conformable to the correct principles of

---

[5] *The Nature and Significance of Economic Science*, 1945 ed., 16.
[6] Joseph Schumpeter, *History of Economic Analysis*, 1954, 681–682.

political economy'.[7] Second, that to the extent this opinion is true, 'there will necessarily be a diminution in the demand for labour, the population will become redundant, and the situation of the labouring classes will be that of distress and poverty'.[8] It was the notion of the population becoming 'redundant' which still strikes fear into those who dread the coming of the robots.

Ricardo did not think that redundancy was inevitable. Higher profits would increase saving, leading to more investment, which in turn would lead to more employment. This meant that 'a portion of people thrown out of work in the first instance would be subsequently employed'; and if, as a result of the increased saving, the same gross produce was produced as before, there need not be any 'redundancy of people'.[9] Nor did he advocate slowing down the rate of mechanisation, 'for if a capital is not allowed to get the greatest net revenue that the use of machinery will afford here, it will be carried abroad, and this must be a much more serious discouragement to the use of labour than the most extensive employment of machinery'.[10]

It is interesting to see that Ricardo's argument against 'slowing down' the speed of mechanisation is echoed in the most recent report on the future of work to have come out of the United States. According to the Council of Foreign Relations, America would lose its 'best and brightest' if it tried to hamper its own technological momentum.[11]

The Subsequent Debate

Let me turn briefly to the economic debate after Ricardo. Central to this was the concept of compensation. The workers made redundant by machines would lose their livelihoods and perhaps even the satisfaction that their work had brought them; but they would be compensated by

---

[7] Ricardo, op.cit., 267.
[8] Ibid., 266.
[9] Ibid., 266.
[10] Ibid., 271.
[11] Council of Foreign Relations, Independent Task Force Report No. 76 on 'The Work Ahead: Machines, Skills, and US Leadership', April 2018.

increased consumption, and the alternative employment which the increased demand for goods and services opened up.

Economists distinguish between 'process innovation' (greater efficiency) which would be 'labour-saving', and 'product innovation', the introduction of a new good or improved quality of goods—which would increase the demand for labour.[12] It is through product innovation that unemployment is prevented from rising. This has been the standard argument.

The post-Ricardian discussion centred on the various forms of compensation which might be relied on to prevent the 'redundancy' of the population:

1. Extra saving out of increased profits leads to extra 'investment'.
2. The general reduction in prices (which, as Ricardo recognised, would follow from process innovation and competition between firms) leads to additional demand for products, triggering product innovation and employment in new lines of production.
3. Initial technological unemployment leads to a reduction in wages, but these reduced wages in turn increase demand for labour and induce a reverse shift back to more labour-intensive methods of production. The theory of wage adjustment remains an important part of contemporary partial and general equilibrium models.

How quickly these compensations come into play depends crucially on how fluid capital and labour are between occupations and regions. The introduction of labour-saving technology will initially decrease the consumption of workers who are made redundant. Unless there is a quick employment response, the cheapening of production will be swamped by a fall in aggregate demand, leading to a rise in unemployment. It was the essence of the Keynesian Revolution to deny the existence of the kind of equilibrating mechanisms postulated above—and certainly their quick operation.

Moreover, even if the delay in the adjustment of demand to supply is only a short-run phenomenon, a series of labour-saving innovations over

---

[12] See Schumpeter (1954: 679, 683–684).

time can add up to create long-term unemployment through a succession of these short-runs. Further, the price adjustment mechanism depends on the general prevalence of competition. If an oligopolistic regime is dominant, then the firm's cost-savings may not translate into lower prices after all.

It is the sluggishness of adjustment to technological shocks that underpins the contemporary view that the benefits of automation are long-run, with 'redundancy' set to rise in the 'transitional period'. But if, as a recent McKinsey Report acknowledges, the transition may last 'for decades', it is hardly surprising that workers are suspicious of this whole slew of compensation arguments.

As we know, Marx denied that any such compensatory processes were at work, either in the short-run or the long-run. The story he told has no happy ending for the workers—at least under capitalism. Under the spur of competition, individual companies are compelled to invest as much of their profits as possible in labour-saving—that is, cost-cutting—equipment. But increased mechanisation doesn't benefit capitalists as a class. There is a temporary advantage for the first mover: 'rushing down on declining average cost curves'[13] and annihilating the weaker firms on the way. But competition rapidly eliminates any temporary super-profit by diffusing the new technology. So the problem of keeping up the profit rate is not solved, only postponed.

The sequence Marx envisaged was this: competition forces mechanisation; mechanisation depresses the average rate of profit (because businesses extract surplus value from humans not machines); restoration of the rate of profit requires an increasingly large 'reserve army of the unemployed'—Ricardo's redundant population. Thus, Marx was able to write that mechanisation 'threw labourers on the pavement'. Marxist unemployment is essentially technologically caused unemployment. The reserve army of the unemployed is temporarily absorbed in bursts of high prosperity, but its longer-term effect is to produce ever rising levels of pauperisation. Thus, for Marx, the sequence was exactly opposite to the classical story: mechanisation might create a febrile prosperity in the short-run, but it would be at the expense of long-run degradation.

---

[13] Schumpeter (1954: 686).

The distributional effects of technical change highlighted by both Barton and Marx have featured prominently in the economists' discussion. Innovation can be labour-saving, capital-saving or 'neutral' depending on its impact on the distribution of income between capital and labour. John Hicks developed the idea of induced innovation.[14] Wage push by workers, threatening the profit rate, induces employers to substitute capital for labour. Most contemporary analysis treats automation as an outside shock, without understanding that it is driven by changes in the relative cost of labour and capital.

The above is the briefest sketch of a complex technical debate. What one can say is that economic theory does not provide a clear answer about the final effect of technological progress on employment. The best conclusion we can salvage from it is that the impact of technology on jobs depends on the balance between process and product innovation, and factors such as the state of demand, the degree of competition, the relation of forces between capital and labour, and the potential for increasing human capital.

Lessons of History

What does history tell us? The Luddites were undoubtedly right about their own trade—handloom weaving. Whereas spinning had been mechanised in factories, weavers were primarily using a handloom which was operated on a domestic basis up until after the Napoleonic wars; in other words, they still owned their means of production. Wages in weaving were very high, as it was a skilled craft. It was not until the early 1840s that the number of power looms in production exceeded the number of handloom weavers.

The introduction of the power loom had three primary results: it concentrated the weaving aspect of cotton production in the factories, it led to the displacement of the handloom weavers, and it destroyed the wages they had received. The displacement of labour from the spread of power looms brought down the wages of the handloom weavers from 23 shillings a week in 1800 to 6 shillings a week in 1830. Nassau Senior advised them

---

[14] In his *Theory of Wages*, 1932.

to 'get out of that branch of production'. They did: 240,000 handloom jobs disappeared between 1829 and 1860 as power looms surged ahead.

What about the larger picture?

Between 1800 and 1915, the population in Europe (minus the Russian empire) grew from 152 million to 315 million. Between 1890 and 1915 (the period of maximum emigration), about 40 million people left Europe for the lightly populated New World. Had this vent for surplus population not existed, Europe might well have experienced some population redundancy in the nineteenth century.

The extra wealth brought by machines enabled voluntary retirement from the labour force, a possibility discounted by the earlier generation of economists, fixated on scarcity. From the mid-1800s, hours of work started to fall from a peak of 60–70 hours a week to about 55 hours by 1914, and then further throughout the twentieth century, as workers took out some of their income gains in greater leisure. The first consequence of wage rises is to increase hours of work in expectation of more consumption; but when people are sufficiently sated they reduce their work effort. Something like this seems to have happened.

Through various mechanisms, then, the nineteenth century did not see a general 'redundancy' of the population. Machinery enabled the economy to support a population that doubled in size at a substantially higher real income. Despite the increase in population, labour's share of gross domestic product (GDP) remained constant throughout the Industrial Age. A degree of redundancy is reflected in emigration figures; a substantially larger degree of voluntary redundancy in reductions in hours of work. Economists were surely entitled to say that such losses as were incurred in the meaning and quality of work and in the more unsettled conditions of life were more than compensated—in the long-run—by the vast gain in material wellbeing. Overall, the Malthusian bogey was obliterated so completely—in the West at least—that the dismal science became the cheerful science.

Relevance for Today

We know that the speed and power of mechanisation has accelerated dramatically. The decisive new advance has been in automated technology and the digital economy, which thrusts mechanisation far deeper into the world of human labour. It is difficult to grasp the enormity of changes going on before our eyes, we don't have enough distance from them. But history and theory can offer some guide.

The theorising provoked by automation is in some respects different from the earlier response to mechanisation. It can be summed up in the notion of 'complements'. This holds that while one effect of automation will be to displace some workers by machines—as in self-driving vehicles—a more powerful effect will be to enlarge the human capacity to complement or work with machines. Thus, the new production processes are best viewed as hybrid, using both human and machine labour. A favourite example is the chess match. A good computer can beat the best human chess player. But the best computer which combines with the best chess player can beat the best computer. Humans will continue to 'add value' to machines. There is no need for them to race against machines, a race they are bound to lose. Rather they will race with machines to an ever more glorious future.[15] This theory, as we can see, replaces the old demand-side story of compensations, with a supply-side story based on the stimulus machinery gives to the development of human capital.

But there are big flaws in the two components of the optimistic view: compensations and complements.

The major weakness of the compensation theory is that it wrongly assumes that meaning in life given by traditional work can be adequately compensated by an increased flow of consumer goods—a typical economistic argument.

The flaw in the theory of complements lies in its vast over-estimation of human capacity. There is no reason why human mental capacity in general should increase at the same rate as machine mental capacity. A minority will be able to race with the machines in the knowledge economy. But a substantial fraction will be 'left behind'. What is to happen to them? Already, the 'left behind' symptoms, and reactions to them, can be seen in increasingly precarious employment, stagnant or even falling wages, and populist protests against both automation and one of its chief agents, globalisation. Even if these distempers are only temporary effects

---

[15] Erik Brynjolfsson and Andrew McAfee, *The Second Machine Age: Work Progress and Prosperity in a Time of Brilliant Technologies*.

of the displacement of labour, optimists themselves concede that the transition period may last decades.

Thus, the idea that a supply shock like automation will automatically set in motion acceptable compensatory demand or complementary supply responses seems to me to be pure delusion. There will, of course, be responses, but they are likely to be highly disruptive, even destructive. Given this, policy must pay much more attention to correlating the rate of change with the capacity of human society to absorb it. This will include slowing down the speed and spread of automation, ensuring its material fruits are equitably distributed, maintaining an adequate level of demand, and providing income guarantees to offset growing precariousness in the job market as robotisation presses wages downwards and eliminates jobs.

This seems to me to be as much as policy can do. None of this, though, addresses the question raised by Patočka: how humans can be enabled to 'feel at home' in world governed by an inhuman and, if not intentionally, inhumane, logic.

References

Barton, J. (1817). *Observations on the Circumstances Which Influence the Conditions of the Labouring Classes of Society*. London: John and Arthur Arch, Cornhill.
Council of Foreign Relations. (2018, April). *Independent Task Force Report No. 76 on 'The Work Ahead: Machines, Skills, and US Leadership'*.
Hicks, J. (1932). *Theory of Wages*. London: Macmillan.
Patočka, J. (1975). *Heretical Essays in the Philosophy of History*. Chicago: Open Court. (1999 ed.).
Ricardo, D. (1817). *Principles of Political Economy and Taxation* (D. Winch, Ed., 3rd ed., 1973). London: John Murray.
Robbins, L. (1932). *The Nature and Significance of Economic Science*. New York: Macmillan and Co. (2nd ed., 1952).
Schumpeter, J. (1954). *History of Economic Analysis*. New York, Abingdon and Oxon: Routledge.
Skidelsky, R. (Ed.). (2015). *John Maynard Keynes: The Essential Keynes*. London: Penguin Classics.

Part I

Work in the Past

3

Patterns and Types of Work in the Past: Part 1

Richard Donkin

The evaluation of work and its many meanings across society, and throughout its evolution, is peculiarly challenging for historians who seek order in the past, interpreting both the outcome of events and the connections between them.

Work is a transforming agency of change, sometimes barely visible in its influence. By some definitions it could be viewed as a simple catalyst of change. We might consider the "work" of the elements—wind, fire and water—in shaping our environment, and the work of all living organisms establishing their relationships, whether in symbiosis or in competition. In the human story, however, work has nurtured both the growth and the flourishing of society. As empires wither, work is the midwife of rebirth and new beginnings. In that sense, work is a fundamental underpinning of the human condition.

Author of *The History of Work* and *The Future of Work*.

R. Donkin (✉)
Cass Business School, London, UK

© The Author(s) 2020
R. Skidelsky, N. Craig (eds.), *Work in the Future*,
https://doi.org/10.1007/978-3-030-21134-9_3

Defining work, therefore, can be as elusive as we choose it to be. It can be as individual as we are ourselves, it can be pliable, broad or narrow, tough or easy, pleasant or taxing, paid or unpaid. We impose our own definitions of work, sometimes seeing it as a joy but often, in a nod to biblical antecedents, we view it as a toil, working as God commanded Adam "in the sweat of thy face".

In mapping the progress of human work we must look deep in to the roots of our family tree, interpreting the uses our ancestors made of their earliest creations, such as the 2,000,000-year-old stone tools found in Tanzania's Olduvai Gorge.

We believe these tools were used for processing food but could they have had another use as weapons? And, if so, which came first? In the film, *2001*, Stanley Kubrick explored man's propensity to violence and the possibility that the sword came before the ploughshare.

Whatever was in the mind of their makers—and we should acknowledge that people were fighting among themselves well before written records began—both tools and weapons were indispensable accessories to hunter-gatherers who established the defining way of life for most of human history.

It is only more recently in our history that we have organised other forms of work in order to deal with settled and expanding societies. In 1966 there was a symposium in Chicago looking at the anthropological studies of Richard Lee among Kalahari bushmen. Marshall Sahlins, a contemporary of Lee's, described hunter-gathering as "the original affluent society".[1] He recognised that hunter-gatherers needed everything around them, but, conversely, they had everything they needed.

In the UK we have examples of early industry in blue stone mining for the making of axe heads in Cumbria and the subsequent discovery of these same axe heads in the Grime's Graves flint mines in East Anglia. This is evidence, not just of industry, but of trade too.

At Boxgrove in West Sussex, archaeologists unearthed a site where a horse had been butchered by hominids some 500,000 years ago. A cliff had collapsed on the site, leaving evidence of flint making, so pristine

---

[1] M. Sahlins, "Notes on the Original Affluent Society", in *Man the Hunter*, ed. R. B. Lee and I. DeVore (New York: Aldine Publishing Company), pp. 85–89. ISBN: 020233032X.

that the outline of someone sitting cross-legged on the ground, napping their flints, was visible among the shavings. We do not know whether the horse they were butchering was killed or whether they were scavenging, but we do know they had great skills. Modern day flint-nappers have ascertained that shaping a tool involves thinking as many as six moves ahead. Considering that a modern chess master would also need to think six moves ahead, was Thomas Hobbes justified in describing the lives of early hunter-gatherers as "nasty, brutish and short"? With what we know now he would need to acknowledge that, however brutal those early societies may have been, their work patterns had already reached remarkable levels of sophistication.

In researching *The History of Work*, I was seeking to answer a persistent question among all those who undertake work: why do we do it? My travels took me to a tribe of aborigines in the Cape York Peninsula in Queensland. I'd been told that the Yir Yoront had no word in their lexicon for work. In fact they do have a word—*woq*—for everyday chores. But this word is never used to describe hunting, their main form of subsistence. Hunting is a joy for these tribes-people (Donkin 2010). Somewhere in our past, the definition of work was transformed from something to accomplish an aim into a more modern definition of something we would rather not be doing. How did that happen when the past two centuries have been awash with labour-saving technologies? Thorstein Veblen's idea of the leisure society was trampled somewhere under the theories of Smith, Marx and Keynes. Work is the protein, interacting with other resources, feeding our seemingly insatiable growth economies that, to our peril, ignore the wellbeing of both the planet and other species.

As internet technologies transform our working patterns through self-organised platforms, often collectively described as Uberisation and the gig economy, there seems to be a yearning for greater understanding of our past, a need for clues that may lead us in to a sustainable future.

One focus of contemporary studies has been the settlements and towns established at a fulcrum of history—the shift from hunter-gathering in to agrarian society capable of creating food surpluses that, in turn, needed finer organisation of work around specialisations such as accounting, planning and management. At sites such as Göbekli Tepe and Çatalhöyük in Turkey there is evidence that a willingness by people to shift from hunter-gathering, where endeavour was centred around the hunt, to a greater and more routine expenditure of sweat in the laying out of streets and construction of buildings, was driven by spiritual needs. Here we have some of the earliest evidence of religious rituals.

In his book, *Against the Grain*, James C Scott has explored beyond the Fertile Crescent in the Levant that's widely recognised as the seat of the agrarian revolution some 10,000 years ago, choosing to focus his studies further to the east into the old Mesopotamia, that area of land between the Tigris and the Euphrates. In the 3000 years predating the agrarian revolution, new cities and towns emerged and often failed after a time, as if people were struggling to stabilise settled centres of population. It seems that all too often, without the necessary anchors of society, people were prepared to walk away (Scott 2017).

Scott is fascinated by anarchistic societies challenging the power of the state. Today we're witnessing an emerging movement for living "off grid" as people choose to pursue less complicated, often less connected lives away from resource-hungry power sources. This movement is tiny but its ability to influence has been magnified among the global audience of the internet.

In our more recent past, the role of religion in the underpinning of working patterns has been crucial. The protestant work ethic, so lauded in capitalism today, was at the very least identified and, in some senses, defined by Victorian thinkers such as Thomas Carlyle, who wrote of "the altar of work". Carlyle influenced Samuel Smiles, whose book, *Self-Help*, chimed perfectly with US puritan ideals popularised in Benjamin Franklin's *Poor Richard's Almanack*. This sense of pulling yourself up by your boot-straps, of triumphing in a journey from rags to riches, depends entirely on a robust work ethic informing the meritocratic ideals at the heart of western democracy. We work as we think.

In the history of work we cannot ignore the role of technology. Work and technology combine in the execution of process. Scott, in his book, refers to fire as an early tool. I view fire, not as a tool, but as an element, like wind and water that could be harnessed to our benefit. We do not know when our ancestors began using fire, but we do know they recognised its potential as far back as 200,000 years ago and possibly much further back than that. They almost certainly *used* fire from natural sources before *making* it themselves from other resources to hand.

The use of fire was a dynamic change, possibly *the* defining change in human evolution that separated man from beast. Once we had harnessed fire, we could evolve ourselves with improving our lives through process. A key process was the cooking of food. That extended our reach, probably not quite as much as bipedalism or the physical attributes of the opposable thumb, but enough to extend our advantage over other species.

The elements again were key in later societal revolutions. Flowing water, working the spinning mills, was the power that triggered the industrial revolution, and it was fire and water, creating steam, that propelled this progression both further and faster. The technologies developed in this period were so profuse that it's easy to forget that they were cumulative—one led to another, almost always driven by need and a recognised purpose. The spinning mills came first to supply the self-employed home-working artisan weavers. Later came the weaving sheds that usurped the very people who had been the earliest beneficiaries of manufactured yarn.

Throughout this change, the organisation and availability of labour was by no means a given. People preferred their semi-agricultural or artisan lifestyles. But the mill offered opportunities for broadening earnings within families.

In the absence of work as a natural state, the early mill owners, men such as Richard Arkwright, founder of Cromford Mills in Derbyshire, needed to provide incentives to lure workers into their factories. People were not drawn to the factories and the dull routines of shift work. These entrepreneurs resorted to advertisements offering houses, even cows and pasture land, close to their mills. It was attractive to the male weaver to have some members of his household as ancillary economic units, so children and women went to work in the mills.

As the mass concentration of employment in industry rolled out across the developing world, new demands emerged for organising and regulating workforces. These forces cannot be divorced entirely from the humanitarian principles that ended the universal acceptance and practice of slavery. The new economics and wage earning incentives of the factory system did, however, offer an alternative to what had hitherto been viewed by slave-owners as unchallengeable custom and practice.

With slavery abolished, the stage had been set for a line of organisational thinkers and managers. To single out a single individual is arguably subjective because this thinking, like the technology in which it is immersed, has a traceable progression of notable contributions. When Frederick Winslow Taylor outlined his ideas on piece-work based on time and motion studies of production workers, he relied on the recent invention of a stopwatch that could measure elapsed time for more than one minute. Without this invention, without his acquisition of such a watch in Switzerland, industry may have waited longer for his "time and motion" studies involving the systematic breaking down of work in to its constituent parts, a practice he called scientific management. Taken in isolation, these insights produced relatively modest gains. Only when scientific management was applied to the moving assembly devised by Henry Ford's engineers for the Model T automotive workshop, did Ford achieve the dramatic production economies that helped to define the consumer society.

The cheaply made and cheaply available yet innovative Model T, made so much profit in its first year of production, that Ford was not only able to return a huge dividend to his investors, he was also able to distribute some of that profit to employees in better wages. He introduced the $5 a day shift at a time when the average industrial wage was about $11 a week. The knock on effect of increasing wages increased the buying power of the working classes and swelled the middle and management classes, leading to the materialistic society we know today.

Today's society, however, is struggling with the economic models that served it so well throughout much of the twentieth century. Economies are creaking under the pressure of globalisation, rampant population growth, dwindling resources, climate change and advancing wealth.

For the first time in our history, collective self-interest globally demands that the needs of the planet over-ride individual desires and ambitions. We need new economics and a new mentality of work, free from the strictures of Protestantism, free from so-called *Bullshit* jobs. Perhaps we need to think more like our hunter-gatherer ancestors of *needs* rather than *wants*. We need a philosophy of work that is connected less to the biblical notion of punishment for past sins, or even to something we see as virtuous. We need work that is positively and intrinsically connected to our environment. We need new thinking about growth and thinking about what we should consider enough in a material sense. Do we need two homes, three cars, a TV in every room and ever more storage for an endless accumulation of stuff? Dealing with all the stuff we make and take from our world makes ever more demands on our precious time. The day, in fact, has arrived that time can be viewed as just as precious, if not more so, than material wealth. To quote the poet W H Davies: "What is this life if, full of care, we have no time to stand and stare?" (Davies 1911)

We need to simplify our definition of work so that it is what we do, nothing more or less. It must also be purposeful work, recognised, if not always by the size of the pay packet, then by its intrinsic rewards and the respect it may earn in oneself and from others. To have any value at all, work must be purposeful. Inputs must be directed at making a better world, a cleaner world and a safer world for our children. Eliminating waste, cleaning the mess we have made across our planet, avoiding excess in everything and focusing on moderation, must be the new focus of work. Anything less is a betrayal of the human legacy.

References

Davies, W. H. (1911). Leisure. Reprinted in Garlick, R., & Mathias, R. (1984). *Anglo-Welsh Poetry, 1480–1980*. Bridgend, Mid Glamorgan: Poetry Wales Press.

Donkin, R. (2010). *The History of Work*. Houndmills, Basingstoke, Hampshire: Palgrave Macmillan.

Sahlins, M. (1968). Notes on the Original Affluent Society. In R. B. Lee & I. DeVore (Eds.), *Man the Hunter* (pp. 85–89). Chicago: Aldine.

Scott, J. C. (2017). *Against the Grain: A Deep History of the Earliest States*. New Haven, CT and London: Yale University Press.

4

Patterns and Types of Work in the Past: Part 2

Richard Sennett

Craftwork seems to be the opposite extreme from high tech. In fact, it can be quite illuminating, both about what technological work could be, and also about the development of skills among workers.

Imagine yourself in the workshop of Stradivarius. How would your career progress from being a young apprentice to becoming a master?

It might seem that you mastered one skill and then you went on to another; progress is all about skill acquisition. In fact, that's only part of the story. In the Stradivarius workshop, you could advance only when you were thrown a new problem for which your prior skills did not prepare you. You had to grapple with the unknown, show you could handle the problematic, in order to advance. For instance, the colour of the varnish—essential to coating any stringed instrument—might seem correlated to the amount of linseed oil in the varnish, because in general the

---

R. Sennett (✉)
New York University, New York, NY, USA

London School of Economics, London, UK
e-mail: r.sennett@lse.ac.uk

darker a stain the thicker it is. But in making a violin varnish, the colour does not equate to viscosity. Composing a varnish is more complicated than the house painter's simple chemistry. The apprentice was thrown into a higher-order, disjointed problem, requiring the re-thinking of "thick" and "thin" in a liquid. Varnish from the Stradivarius studio is difficult to replicate today precisely because the varnish formulae became progressively non-intuitively obvious.

In this example, skill is not additive. One thing does not teach you how to do the next. Knowing how to colour, in high-level varnishing, in fact poses a puzzle about viscosity: there is a dialectic between problem solving and problem finding. That is absolutely key to developing a skill more generally. Dealing with an unforeseen problem or difficulty is more important than "applying" existing knowledge. Or to use another cliché, "skills transfer" does advance the development of higher-order skills.

What is the relationship between hand and head in moving from established skills to establishing new skills? To take an example from our time, in my design studio at Harvard, we use both CAD—computer assisted design—and tools called pencil and paper. It was thought originally that CAD, particularly the relatively sophisticated sort that we use, would replace drawing by hand: all you needed to do was plot points on the screen and instantly you could produce a design. What we've found during the last 15 or so years is that designs produced on screen are much less interesting and innovative than the designs drawn by hand. Why?

A traditional answer, given by Ruskin in the nineteenth century, was that hand work puts the craftsman more "organically" in touch with his or her labours. This is too Romantic an answer. We worked for a couple of years with people at the Harvard Medical School to try to find out why it is that hand drawing can be more innovative than CAD design. Their argument to us was instead that the sheer difficulty of drawing by hand, and even more, the uncertainty of how the hand will move, serves as a stimulus to innovation. Again, resistance and difficulty were breeding skill, in this case an imaginative skill. Researchers like Frank Wilson have tried to validate neurologically this connection between the head and the hand: the hand problematises the head.

So it's not a matter of being a Ruskinian Romantic, nor a Luddite bent on smashing the infernal machine. The architectural designer needs both CAD and the pencil—the pencil to think, CAD to execute. So is this the case with 3D printing: so easy to use, which means so numbing. Good 3D modellers start with thinking in wet clay. The uncertainty and open-endedness of actual physical effort is a way to open up what should be produced as an object, which you can then summon into being by 3D printing via mimetic algorithms on screen.

Rather than craftsmanship, particularly hand-craftsmanship, being put out of date by any kind of mechanical labour, we have to have a much more sophisticated notion of connecting the bodily work we do to the mental labour we do. That is not connection that is made in most discussions about high tech today. It just assumes a replacement, as though you could dismiss the body and therefore dismiss the problems of difficulty or imagination, from the world of work.

I would like to make an observation, which follows from this, about de-skilling. In Adam Smith's discussion of the division of labour, he understood intuitively that a machine, organised in a certain way, could "de-skill" by stupefying the worker. The key part of his observation was the modifier, "in a certain way". This is a matter of how machines themselves are configured. Configured in one way high tech machines do indeed stupefy and so de-skill the worker. The most obvious example is Google Maps, which replace environmental reasoning and place memory by prescriptive routing, involving no perceptual intake on the part of the user. But the problem of the stupefying machine is more generally inscribed into "user-friendly" technology. You need not think about how your tool works, what problems or possibilities it excludes in the name of friendliness. You do not need to worry, and so you do not think.

The challenge in labour, it seems to me, is how to organise technology so that it is a partner rather than a replacement. What we are trying to do in our design lab, as is the Media Lab at Massachusetts Institute of Technology (MIT), in experimenting with computerised prostheses in surgery, is to use devices like CAD or 3D printing as technologies which enable more human choice, rather than replace choice with one standard. As in Stradivarius' studio, we are trying to design these two machines so that they are challenging rather than "friendly".

As a generality, I would say that thinking fresh about labour requires new ways of connecting material and sensate experience to mental understanding. That means recovering the lessons of craftsmanship in the age before machines dominated the productive process—which is my defence of craftsmanship to you.

5

Patterns and Types of Work in the Past: Wageworker and Housewife from a Global Perspective: Birth, Variations and Limits of the Modern Couple

Andrea Komlosy

The modern couple consists of a breadwinning husband and a wife who lives from his earnings. He goes out to work; she stays at home and takes care of the family. This may be a fiction, hardly complying with the reality of most families. It has nevertheless been a successful model, becoming the prototype of matrimonial relationship not only in Western Europe but all over the world, since it was given birth with the onset of the Industrial Revolution, exercising pressure on its counterpart, the family economy that has been the predominant mode of working and living before.

Author of *Work: The Last 1000 Years*.

A. Komlosy (✉)
Institute for Economic and Social History, University of Vienna, Vienna, Austria
e-mail: andrea.komlosy@univie.ac.at

The family economy is a household economy, presided by a husband and a wife. They each have their place and their obligations in the works, carried out in the household. Households range from nobles to ordinary people. Think of a peasant couple who deliver crops for the market and the manorial lord and also produce food, shelter, clothes and other everyday items for their personal use. Or think of an artisan couple, who carry out a specialized profession, like weaving or metalworks, both working at home: the master craftsman with his apprentices, the lady of the house supplying lodging, food and care for kin and non-kin members of the household. Depending on place, time and craft there is a clear gender division of labour: usually the husband is the power-head of the whole household enterprise, which includes children, non-married relatives, servants, apprentices or labourers. It goes without saying, however, that female and male operations, both for the market as well as for immediate use and consumption, are considered work.

The transition from a household-based economy to a workplace economy shows a broad variety of patterns. It also differs in time and place. The new type of couple that came along with the relocation of work to places outside of the living place started in eighteenth century Western and Central Europe. It had predecessors, however. Some trades were carried out far from the home and required workers to leave the household, for example construction works or road and transportation works. Mining always required a separation of living and working place. But nobody thought of denying work-character to what was carried out by family members in the households back home, either for subsistence, market or wages. Many itinerant trades also relied on the cooperation of the family members. Moreover, the transition from home to outside workplace spread slowly at the beginning, only involving some trades and products, while the majority remained rooted in households.

There are several reasons, why the modern couple did not become the dominant family model immediately after its first rise in Western industrial regions. A sudden change only took place in some industrial branches, like cotton spinning and weaving, where factories, power-driven machines and wagework took over quickly at the turn from the eighteenth to the nineteenth century, while the rest of the economy continued being family-based. Where these branches widely spread, the new pattern broke through. Transitions differed depending on class, branch, town or countryside.

The ideological foundation of the male breadwinner concept was laid in bourgeois families of factory owners, whose wives were not supposed to join husbands in professional work anymore, as they had done in household craft industries. But far from gaining freedom, the work they carried out in the homes, supporting husbands and children, organizing social life and overseeing the servants, was not considered work anymore. Only paid work was defined as work from now on. The strenuous parts of the housework were done by servants whose status also changed. On the one hand they were part of the bourgeois household, where they lived and worked, on the other hand they faced a professionalization of home services, becoming wageworkers. On these grounds, the illusion of the non-working lady of the home could be claimed.

Basically, the modern couple was a phenomenon of industrially developing core regions in Western Europe and North America, spreading along with the process of industrialization. There is a broad assumption that it took over, when industrial transformation reached peripheral regions and the non-European world. I have deep doubts about this diffusionism. Did it ever become a generalized pattern? Did it reach peripheral regions at all?

We can observe a non-synchronicity in introducing factory-jobs and the taking over of modern kin-family patterns. There is no linear tendency of development, however. We see hesitation, retardation, even resistance of family households to recede and transform. Peasant agriculture and crafts households resisted factory or cash-crop competition, carrying on the household economy and their working characters, husband and wife, into the factory or agro-industrial age. While in some Western European countries craft workshops and peasants were replaced by factories and big farms pretty soon, others resist until today, when we even see a revival of family agriculture and craftsmanship. So instead of a replacement, old and new working characters competed, overlapped and merged into new hybrid forms, giving way to the following archetypes:

1. The male breadwinner/non-working housewife couple, supported by servant home service workers.
2. The middle-class strenuous copy, where the husband's income does not allow for servants, so that the burden of housework all rests on the housewife's shoulders.
3. The proletarian reality, obliging both husband and wife to do wage work. Female wages are below male ones, because women's paid work is considered only as an additional income, while her main designation is that of a housewife and mother. She carries the double burden of wage work and housework. Also working-class men complement wageworking hours by home and subsistence work.
4. The double burden explains why many working-class women dream of achieving the position of a housewife-only, delegating wage work to her husband, hence catching up with the middle-class model.
5. Last but not least, peasant and craft households, who are able to maintain a cooperative household economy, often rely on additional gainful occupations to complement business revenues. If there are no jobs available where they live, some family members become temporary or permanent migrant workers. They can either stabilize the household economy by their remittances or return investments, or they form a bridge for further family members to follow migration chains and build up modern couple families at their new destination.

The following sections show the varieties of implementation of the breadwinner/wageworker-housewife couple in various parts of the world. They depart from an industrial core region in Central Europe, Lower Austria, in the nineteenth century, where these developments had a class-encompassing character, before contrasting metropolitan developments with those in the colonial world, where the household economy either resisted or was destroyed without reproducing the Western family pattern. Finally we discuss the limits to the introduction of the modern couple in the postcolonial world in the second half of the twentieth century.

Central European Core: Lower Austria

Lower Austria is among the most developed early industrial regions ("lands") of the Habsburg Monarchy, surrounding the capital of Vienna—a high ranking place of industrial production and consumption (Komlosy 2010). Contemporary observers seemed impressed by the density of cotton mills in the region south of Vienna "resembling English industrial districts" (Blumenbach 1834).

No wonder that factory work was spreading rapidly between 1800, when the first cotton mills were set up, and 1900, when a fully integrated industrial landscape had developed. We investigate the conditions under which the modern couple became the dominant family pattern in the region. This was first and foremost the case among the bourgeois factory owners. Apart from some associations of merchant and noble capital owners, whose factories were administered by managers, most factories were owned by individual families. Due to good water supply, villages were transformed into factory locations, typically comprising production buildings, housing for workers, owners' villas as well as shop, school, pub or assembly hall for events, provided by owners or communities. Owners' families often split their time between the factory site and an urban residence, most prominently in the metropolis. Family management and other family members often commuted between town and factory village, multiplying the household activities for the lady and the servants. It was self-understood that the lady was not supposed to have a profession and her education only aimed at skills needed for representation. Some ladies were successful in becoming organizers of famous society events, building up on the solid financial basis of their "breadwinning" or profit-making husband.

There existed a broad variety of working-class families. When workers happened to own a house at the place where they performed factory work, they can be classified semi-proletarian—with one foot in wage work and another one in small agriculture for sale or subsistence. Seen from the peasant household, taking a waged occupation was a means for the family agriculture, to which all household members contributed work, to survive against the competing pressure of agro-business and cheap crop imports. Often peasant-workers had to migrate to find a job, as agricultural labourer, servant, construction or industrial worker. As long as workers returned on a daily, seasonal or temporal basis, they could still consider themselves as well as the family members who worked the plot, peasants. Whoever was excluded from heritage or gave up the peasant land, became a landless worker.

Similar to English or Scottish factory villages, like New Lanark or Saltaire, the early Lower Austrian textile mills recruited single men, women and children from rural regions, building separate hostels for male and female workers, who worked and lived in the factory, where they received (for pay) shelter, food and some basic education for childworkers. Working days and burdens were so hard that people died before they reached their 30s, and were replaced by a younger cohort. Even if some capitalists opposed, factory owners soon realized that workers needed families in order to survive. A single proletarian did not allow for reproduction. This is why industrialists soon built workers homes—cheap flats for the ordinary workers and modest individual homes for the skilled ones. They also provided basic medical aid and social support in case of disease, before public health insurance was introduced from the 1880s onwards. Workers' colonies became home to worker-housewife couples. Women were occupied, but not paid for keeping the household going, with labour-intensive activities like gardening, cooking, washing or repairing, keeping them busy all day. If more wage income was needed, wives had to take up additional employment. The low wage was justified by the primary female dedication for family work, while wage work was considered a temporary extra. Certain industries, like knitting, lace making or sewing preferred to outsource work to workers' homes, where mothers could more easily combine housework and contract work. Homework can be found in distant rural regions as well as in crowded urban suburbs, where putting out systems flourished especially in the garment and leather industries. It cannot astonish that these women aspired liberation from work, hoping that moving-up into the middle-class and relying on their husbands' wage would improve their situation. In the case of skilled workers, wages were sometimes high enough to feed a family, so that the male breadwinner model could enter proletarian households. It did not bring an end to drudgery, however, especially as social advancement usually increased expenses and housewives had to compensate for lack of money by extending the scope of their unpaid activities. The images communicated in the communal housing architecture of the interwar "Red Vienna", social-democratic city-government clearly demonstrate how the bourgeois understanding of family roles entered labour party and movement. On numerous walls, statues and reliefs you can see mothers with children, in contrast to the presentation of men as architects or workers.

A look at lower middle-class families clearly shows that housewives had long, hard, exhausting work-days, often lacking help, company and recognition. After all, they also carried a double burden: on the one hand they acted like bourgeois ladies, while on the other hand they did the work affluent middle-class ladies delegated to their paid servants. Both husband and wife identified with the roles provided by the male breadwinner model. Even if husbands had low salaries, they gained self-respect from not being dependent on their wives' income. Marriage laws entitled husbands to decide whether wives could enrol in a labour contract and they made use of that, even if emancipated wives wished to work for money. Most women internalized their role as housewife and mother just like their husbands, voluntarily fulfilling a profession which was denied the status of work. Asked if they work, the classical answer was: "No, I am a housewife". You can still hear it today. As they follow standards of fashion, representation and cultural life set up by upper classes, these housewives work hard to compensate lack of fortune and property, trying to correspond to an image they cannot afford. Trying to pave a better way for their children by urging them to high-flying performance often contributes to tensions between the partners and generations.

Radical women and men have questioned bourgeois gender roles and the division of labour between the sexes since the French Revolution, and the feminist movements in the second half of the nineteenth century achieved improvements in women's rights and mentality. It was not until the second feminist wave of the 1960s that legal equality between men and women in Western societies was implemented; in state socialist Eastern Europe equality was extended to the right to work, which did not allow women to be mere housewives. In East and West more and more women entered professional gainful employment, and the breadwinner-housewife couple lost grounds. Women's primary responsibility for family affairs, while men were defined by profession and career, did not change. Therefore many women worked part-time. Non-paid housework continued being excluded from work and value-creation. Only when the regular work regimes in gainful employment gave way to flexibilization, precarization and new combinations of paid and unpaid activities in one's work-day, did the modern understanding of what can be considered work become fuzzy.

Colonial India

Indian states and regions had been among the most advanced industrial regions of the world until the eighteenth century. Historical statistics attribute 24.5% of global industrial production to India in 1750 (Bairoch 1982). We must not think of factories, but of putting out systems made up by merchants, contracting rural households to deliver yarn and plain, painted or printed cloth. The English East India Company entered local and regional commodity chains as early as the seventeenth century, redirecting Indian textiles to export markets in Europe, Africa and the Americas. Comparative studies about living standards revealed that rural textile producers enjoyed quite a good life. Because of the embeddedness of rural crafts in subsistence agriculture living costs were low, as were wages, rendering the produce highly competitive. In the course of the eighteenth century British interests moved from finished textile imports to raw cotton and other raw materials to be manufactured in British factories. Transforming industrial producers into unfinished cash-crop suppliers took a while. The family economy, with its specific regional products and patterns of status, professional and gender division of labour, strongly resisted, even when British and other Western European nations closed their markets to Indian calicos from the 1700s onwards. Sven Beckert (2014) depicts the British endeavours to turn India, where British colonial interests were represented by the East Indian Company from 1757 onwards, into a supplier of raw cotton. They did not succeed, because peasant craft-families did not want to give up their diversified, mixed agro-industrial local economy (Komlosy 2018).

It was much easier to expand cotton plantation systems in the Caribbean and the US South, where European mill owners turned to at the turn of the eighteenth to the nineteenth century, because slaves lacked the means of resistance, while Native American resistance was met by their removal and repression.

The break-up of the Indian economy, relying on a mix of agriculture and home industries, subsistence and market production, required a more sophisticated procedure. While finished Indian textiles were closed off from export markets by protectionist custom regimes, new taxation systems put pressure on peasants to adapt to British demands. In Bengal, the "Company state", the "Permanent Settlement Act" (1794) did not only introduce Western property relations; it also introduced a land tax system, which drove mixed household economies into cash-crop specialization, mainly opium, cotton and indigo, if they wanted to survive. It caused a hunger crisis, to which millions of people fell victim. Family households that could not sustain themselves were forced to migrate to tea plantations that were set up in Assam or Darjeeling or to distant British plantations in the Caribbean, Pacific and the Indian Ocean. There, the British ban on the slave-trade (1807), followed by the abolition of slavery in their colonies (1833) had left a void in plantation labour, that was filled with so-called Indian "coolies"—predominantly young male workers, formally free, but tied to their employers by long-term indenture contracts in exchange for travel costs. Like slave plantations, coolie plantations relied on single men, and it was only after the end of the indenture that families and households were founded, leading to an Indian diaspora in the British colonies. If former coolies were successful in setting up their own businesses, they followed the pattern of the family economy that relied on family work and family networks.

Indian textile manufactures were ousted from global markets, contributing to a sharp decline in industrial output. Under the British Raj the Indian demand for textiles was met by imports from Britain (Western style textiles), while everyday cloth was produced by small workshops, based on the household economy. With some exceptions in Gujarat, Bengal and Madras, factories were only introduced in the twentieth century. Wageworker-housewife couples therefore remained a domain of colonial administrators, who imported the model from the West, and by those urban Indian middle-class citizens, who adapted their lifestyle to Western standards. Local craft workshops could resist the competitive pressures of Western imports by cheap supply of labour. Compared to the eighteenth century they were marginal, however. India's share in world industrial production fell from 24.5% in 1750 to 8.6% in 1860 and 1.7% in 1900 (Bairoch 1982).

It follows that there was no place for modern wageworker-housewife couples in colonial India apart from in the colonial administration. Instead, every activity that was performed or pooled in the family household and contributed to the survival of the family was valuable. Wageworkers had to be peasants, housewives had to be textile workers; those who could not be fed had to leave and became workers on plantations or in the construction of railways, ports and urban infrastructure which expanded under British rule. In all these sectors there was a clear hierarchy of activities and professions according to the system of stratification along birth (*jati*), descent and professional (*varna*) lines that was transformed into the rigid cast order under British rule (Mann 2005). *Varna* and *jati* affiliations reflected a hierarchical order of the society, they did not classify between market and non-market activities, however.

Family, Couple, Single or Composite Households?

The two cases of historical textile regions show a sharp contrast. Lower Austria is among the birth-places of the modern couple. However, it demonstrates big variations according to class, branch and region. The modern couple, consisting of breadwinner/wageworker and non-working housewife is rather an idealistic construct than a practical model. In the social praxis, it has to be appropriated according to the situation. When women took up professional gainful occupations in the twentieth century, it gave way to a double breadwinner-one housewife household.

Colonial India is too big and too diverse to allow general conclusions. We can nevertheless state that the Western model did not enter the social life of the majority. Statistics speak a clear language reflecting the destruction of the Indian economy, based on exports producing within the household economy. Instead of giving way to a rising proletarian class of wageworker and housewife couples the family households adapted to the new situation. If they could not maintain their means of production, that is land, farmhouses, workshops, devices, they transformed into income-pooling households. This is not only true for India, but applies to global peripheries in general.

In the global South the modern couple did not gain momentum. It was only attractive for postcolonial upper and upper-middle classes. They copied the model that was developed by European industrialists, free professions and privileged private or state employees: A model in which neither master nor lady of the house had to do housework, because they employed servants. Today's global old and nouveaux rich are equally used to share their big homes with highly dependent service personnel. From Brazil to Delhi, only a small number of middle-class couples live in nuclear families, in which both partners have gainful occupations and do not have servants to do the housework.

Lower and lower-middle classes in the global South live in income-pooling family households; in many cases they have a trans-regional character, including family members in the countryside (eventually working the land), family members in local towns (eventually working as factory-workers, in administration or in the informal economy), as well as family members who emigrate to work abroad. This description may serve as a prototype which is adapted by each family in a specific way. Occupations and responsibilities change along the life-cycle and often single family members get lost, not contributing to the common household anymore. Such a household can ideally unite several breadwinners, wageworkers, peasants, migrants, housewives and househusbands, not to speak of those who do small jobs and informal activities just to get along. There is no exclusive couple, however, tied together by wage work and housework.

Both the couple and the trans-regional household face erosion. The trend towards singling and individualization affects all sorts of families. Couples divorce, income-pooling households get uprooted by poverty and migration. What is the future likely to bring?

Complex family households, based in agriculture, manufacture and social reproduction, as we know them from pre-industrial and pre-colonial periods, cannot be brought back. Maybe we do not wish them back for their patriarchal structures and hierarchies. How could they be replaced in modern societies, overcoming the gender and working roles of the nuclear family household?

Future creative work-life arrangements could learn from the inclusive assessment of work in the family economy, including commodified and reciprocal activities, not forgetting the fiscal contribution of gainful work to finance public expenditures. It goes without saying that such arrangements would require a different distribution of gainful work, house and care work, public work and learning during one's lifetime and everybody should be allowed and obliged to participate in all of them. Like in pre-modern households, family could be interpreted in a broader understanding, including kin and non-kin members, while overcoming patriarchal gender divisions of behaviour, education, profession and public roles.

In the commodified sector, family-based enterprise could be supported or replaced by community-based enterprise, practising self-management, co-working and co-living in different forms, taking account of the changing needs and preferences from young to old. Cooperations would primarily aim at neighbourhood and regional commodity flows, complemented by trans-regional cooperations wherever wishful. A main branch where community-based enterprise can be more attractive than the nuclear family is care and social work in all its manifestations. It would equally be useful to include care, social activities and leisure into the work-day organization of commercial enterprises.

What is carried out in a reciprocal way depends on the organization of the commodified sector. Redefining the relationship between profit and commitment would allow carrying out basic reciprocal operations within a commercial private or public enterprise. A whole range of activities should rather stay outside the monetary realm, following social, moral and gift systems of exchange, and be located in families, households, neighbourhoods and networks.

Various concepts of new future work-life arrangements have been developed elsewhere in greater detail. Reminding them here is a call to make use of historical concepts in order to shape the future.

References

Bairoch, P. (1982). International Industrialization Levels from 1750 to 1980. *The Journal of European Economic History, 11*(2), 269–334.

Beckert, S. (2014). *The Empire of Cotton. A Global History*. New York: Alfred Knopf.

Blumenbach, W. C. W. (1834/1835). *Neueste Landeskunde von Oesterreich unter der Enns*. Güns.

Komlosy, A. (2010). Austria and Czechoslovakia. In L. H. van Voss, E. Hiemstra, & E. van Nederveen Meerkerk (Eds.), *The Ashgate Companion to the History of Textile Workers, 1650–2000* (pp. 43–73). Farnham: Ashgate.

Komlosy, A. (2018). *Work. The Last 1000 Years*. London and New York: Verso.

Mann, M. (2005). *Geschichte Indiens. Vom 18. bis zum 21. Jahrhundert*. Paderborn: Schöningh.

Part II

Attitudes to Work

6

Attitudes to Work and the Future of Work: The View from Economics

David A. Spencer

Introduction

Work is an obligation to the person without the means to live without it. It is a compulsion—an activity that most of us cannot avoid. Under capitalism, work is performed to secure income. Wage-labour is not chosen in this sense, but rather is imposed by the system of work that prevails in capitalist society. Like death and taxes, work is something we must face, whether we like it or not.

Yet, while an unavoidable necessity, work has important impacts upon us as people. It shapes who we are and are able to become. People are as much the products of their work as the things they produce and the services they deliver. Work can add to the quality of human life, by creating opportunities for skill development, positive social interaction and personal achievement. Equally, it can be degrading and alienating in ways that damage the well-being and health of people. Society condemns

D. A. Spencer (✉)
Leeds University Business School, Leeds, UK
e-mail: D.A.Spencer@lubs.leeds.ac.uk

© The Author(s) 2020
R. Skidelsky, N. Craig (eds.), *Work in the Future*,
https://doi.org/10.1007/978-3-030-21134-9_6

sweat-shops not just because they pay low wages but also because they feature oppressive and life-limiting work conditions. Demands for workers' rights speak to a deeper need for work that fits us as human beings.

In this chapter, I consider the contribution of economics to the study of work. I show how economics has seen work as a means only. Where it has considered the welfare effects of work, it has portrayed work as a 'bad thing'. Economists have regarded work as something that people do for extrinsic reasons and that they would avoid if they could. I contrast this portrayal of work with rival perspectives found in non-mainstream or heterodox economic thought. These perspectives take into account the potential benefits of work itself and consider the possibility of achieving a state where work brings meaning to human life. I also consider critically the application of economics to the study of automation and the future of work. My argument will be that the economics of work needs to be radically reworked in order to understand the nature of work, both now and in possible futures to come.

Work as 'Bad'

Economics has been dominated by a particular conception of work. This conception has proved remarkably persistent, resisting change through time (Spencer 2009). It associates work with unpleasant activity. In the language of economics, work is a 'disutility': an undesirable chore that is resisted by humans. Only the allure of income, it is argued, motivates humans to work and without this allure no work would get done.

The view of work as 'bad' has gone through different stages of development in the history of economic thought. An early view linked to classical economics saw work as intrinsically painful. Adam Smith (1976: vol. 1, p. 47) referred generally and indiscriminately to work as all 'toil and trouble'—for him, like all other classical economists, the activity of work evoked displeasure in human beings and it was only because of the pull of higher income that humans engaged in work. The irksomeness of work was to be contrasted in all cases with the desirability of a non-work state of ease or idleness.

A second view stemming from neoclassical economics has placed emphasis on the lost opportunity for leisure time. This view has since come to dominate in economic textbooks. It assumes that work is resisted not because it causes pain but rather because it leads to the denial of the opportunity to spend time as leisure. People, it is assumed, derive utility from leisure time and will only give up leisure hours if compensated by wages. Here the direct cost of work is not considered; rather the opportunity cost of work time is emphasised in the definition of the human resistance to work.

A third view has evolved out of the economic analysis of agency and sees workers as 'lazy' and incorrigible 'shirkers' (see Jensen and Meckling 1976; Williamson 1985). Workers do not resist work because they find it painful in itself or because they 'love' leisure—rather they resist work because they have an innate desire to avoid work. The assumed inherent laziness of workers is used to justify measures that enforce work via the provision of hard incentives, close monitoring, hierarchy, and the threat of contract termination. Employers, it is assumed, must cajole and coerce workers to work.

These three views are open to question. The first assumes erroneously that the burden of work is pre-given and immutable—it fails to capture how workers are led to resist work due to its organisation. Adam Smith, to be sure, recognised and condemned the dehumanising effects of repetitive work (Smith 1976: vol. 2, p. 782). He worried, in particular, about the loss of workers' intelligence through their performing the same tasks repeatedly. Yet, at the same time, Smith gave the impression that work would be costly without the detailed division of labour. There was a resignation to work's cost that obscured its roots in the organisation of work. By extension, Smith, with other classical economists, missed the scope to turn work into a source of pleasure. In classical economics, the pain of work was to be accepted in the pursuit of higher growth and higher living standards.

The second view linked to neoclassical economics fails to address the lives of workers at work. The work-decision is reduced to a simple trade-off between two goods, namely income and leisure time. Work figures in this formulation as a means to income only. There is no attention to the way in which people are shaped by the work they do. The fact that workers may embrace work for its own ends or suffer deprivation in the act of working is ignored. There is also a neglect of the capacity to improve the quality of work and to realise a state where workers are able to derive benefit from work itself.

The third view implies falsely that workers' lethargy is the main cause of conflict at work. In effect, work resistance is viewed as a product of workers' genes, not of the way that work is organised. There is a failure to see how employers can manufacture conflict through their own behaviour and how resistance to work can arise from workers' exposure to a harsh and oppressive work environment. At worse, this view casts an ideological shadow, targeting workers for criticism and portraying employers as benevolent actors who implement the 'right' mechanisms to prevent shirking by workers. The claim that economic theory is neutral, in this case, is undermined.

Underlying the above three versions of the work as 'bad' thesis is the notion of homo economicus as an idler and sloth. Humans are deemed to crave a life without work. Nirvana for the worker depicted in economics textbooks is a state of luxurious leisure, where consumption is maximised and work is minimised. This view is not just biased in its focus (i.e. it says nothing about the qualitative content of work); it also lacks any awareness of the need that humans have for productive activity. As we shall see below, we are required to look beyond conventional economics for an understanding of the broader significance of work, both as an activity and an influence on human well-being.

Work as a Possible 'Good Thing'

Several different contributions from within heterodox economics have examined the notion of work. Here I pick out two key contributions, namely those of Marx and Veblen. These contributions ask us to think differently not just about work itself but also about human nature and the goals of progress.

In the writings of Marx, work was viewed as more than just a way to earn a living; it was also viewed as a potentially liberating and life-enhancing activity. Marx (1977) described work as a part of humanity's 'species being'. Through work, humans could express their creativity and achieve self-realisation (Sayers 2005). Work had meaning and importance in its own right and could be a way for people to develop and progress as creative beings. Indeed the task was to create conditions where work was fulfilling.

Yet, while stressing the latent positive potential of work, Marx stressed the alienation of workers under capitalism. Work represented a forced activity in capitalist society—it was performed because of the need to earn wages. Wage-labourers were disempowered by the wage-labour relation and when at work faced having to take orders from capitalist employers. The alienating aspects of capitalist work were evidenced by the fact that workers looked upon work as a purely instrumental activity. Workers shunned work not because they were lazy but rather because they confronted the unfreedom of wage-labour. Marx stressed how work resistance had a specific meaning under capitalism—it was not universal in this sense, but rather linked to the power imbalance at the heart of the capitalist employment relation.

The problem of 'alienated labour', however, could be resolved. By transcending capitalism, a non-alienating form of work could be realised. For Marx, work could and should be recreated as a positive activity. This entailed no less than a revolution in society and the move to communism. In a future communist society, the reduction of work time through the use of technology and the creation of communal forms of ownership would enable work to be enjoyed in the same way as activities performed outside of work.

Veblen (1898), from a radically different standpoint, argued that humans were not natural idlers, but rather were creators with an 'instinct of workmanship'. Resisting the neoclassical view of homo economicus, Veblen showed how humans would embrace efficient and skilful work. Humans had progressed by devotion to work and the possibilities for future progress depended on peoples' continued pursuit of 'useful effort'.

To be sure, there was resistance to work in society. But this resistance reflected not on the costs of work itself, but rather on the effects of a 'pecuniary culture' that gave primacy to earning and spending money. Veblen bemoaned how the 'instinct of workmanship' had been crowded out by the dominance of pecuniary values in society. Under capitalism, in particular, there had developed a cultural aversion to work based on the value of money-getting and the effect of this aversion was the eclipse of the merits of productive activity.

In Veblen's view, there was a need to challenge the dominant pecuniary culture of capitalism. Veblen did not side with Marx's call for communism, but instead signalled the benefits of a move to different economic and social circumstances where the 'instinct of workmanship' could be more fully realised. While vague on the details, Veblen saw a key role for skilled occupations (notably engineers) in the creation of a better world where work would be valued for its own ends, not just for the money it brings (see Layton 1962).

The ideas of Marx and Veblen have had a much greater impact on research outside of economics. Although their ideas have offered direct criticisms of the economics of work, these ideas have been ignored by economists. As a result, the narrow view of work as 'bad' has persisted in economics, despite its many faults.

Here three key insights can be gleaned from the competing perspectives of Marx and Veblen. The first is the potential for work to be rewarding in itself. Work is not an unending drudge but rather it can be potentially uplifting. Here there is a rejection of the idea that workers are congenitally averse to work—rather there is emphasis on the importance of work in meeting human needs for creative action. Secondly, there is a stress on the specificity of the costs of work. Workers are led to resist because of the economic and social conditions they work and live under. In the case of Marx, as suggested above, there is a focus on the problem of alienation as it exists under capitalism. For Veblen, again as mentioned above, attention is given to the importance of culture in creating an aversion to work among people. Thirdly, it is emphasised how the costs of work can be negated. Here we are faced with conflicting paths—from a 'soviet of engineers' in the work of Veblen to the communist society favoured by Marx. Whatever position is taken, however, there is stress on the possibility and indeed necessity of returning meaning to work and creating the conditions for people to flourish in the places they work. The goal of elevating work's quality, as discussed further below, has direct relevance for discussions of automation and the future of work.

The Promise of Automation?

The possibility for mass automation and with it the disappearance of work has been discussed for many years (Mokyr et al. 2015). Concerns have been expressed about the prospect for machines to replace humans, giving rise to higher unemployment. Luddite-like anxiety has been fuelled by a fear of a future where jobs are scarce in number and where poverty levels increase significantly. Yet, by contrast, there has been the hope that automation processes will deliver a better future where human freedom is enlarged. Indeed, some writers have championed automation as a route to a superior 'post-work' society (Gorz 1985).

Such concerns and hopes have resurfaced in the present, due to predictions of mass job losses via automation (see Spencer 2018). The evolution of machine learning and artificial intelligence, it is claimed, will allow for the replacement of human workers across myriad jobs. Pessimists, like in the past, worry about how society will adjust to a world without work (Ford 2015). Optimists, reviving the older visionary perspective of Marx, embrace 'full automation' in the move to a state of luxury consumption, where work is absent (Srnicek and Williams 2015).

Here two themes can be addressed. One is the place of economics in the above debates. From the perspective of economics, automation sounds few alarm bells. On the one hand, there is the view that new jobs will be created to offset those lost by technological progress (see Autor 2015). Economic theory tells us that jobs will be created as technology depresses prices, adds to the number of available products and boosts real incomes. On the other hand, even assuming net job losses, the prospects remain positive, on the basis that individuals can adjust their labour supply to accommodate the reduction in work. Notably in any such adjustment, workers are seen to suffer no loss from the decline in work but rather are seen to 'choose' more leisure, in accordance with their (given) preferences.

The above view is evidently problematic. The most obvious problem concerns the assumption of 'free choice'. It is as if workers can allocate their time between work and leisure as they please. The above assumption is a fiction invented by economics. In the real world, workers face a necessity to work. Workers need to work, even if wages are low and the quality of work is poor. To assume that workers can somehow 'optimise' in the face of automation is to paint a false view of the world in which we live.

There is also the problem that the loss of work is seen to inflict no harm on workers. Assuming workers can compensate for the loss of income by working less, they will always benefit from the reduction of work. This abstracts from the cost incurred by the loss of work itself. There is now a well-established literature that confirms the non-pecuniary penalty of unemployment (see e.g. Clark and Oswald 1994)—the costs of unemployment extend to the loss of opportunity to perform valued skills and to socialise. This literature points to the fact that people value work for its own sake and how the elimination of work may bring psychological and social costs. While economists persist with the view that work is always a 'bad', they will miss the above costs and in turn will underestimate the potential welfare impacts of automation.

The second theme to be addressed here concerns the objectives of automation. In economics, the objective of automation is to reduce work and increase leisure—the ideal is to create a situation where people can consume to their heart's desire with no work. This is again based on the notion that work is a disutility and leisure is a good. Such reasoning, however, fails to see how automation can be used not just to extend free time but also to create more satisfying activity, including in the work domain.

Much debate has focused on automation as a route to greater freedom outside of work. Keynes, notably, embraced automation as a way to reduce work hours. In an essay published in 1930, he thought it would take 100 years to achieve a 15-hour working week (see also Skidelsky and Skidelsky 2012). Keynes, in line with standard economic theory, represented work as a disutility. Automation was to be encouraged, in this case, to provide more leisure time for people to pursue activities of their own choosing. Again the goal was to minimise the human exposure to the disutility of work.

Keynes differed from his neoclassical colleagues by seeing in leisure the potential for creative activity. He rejected the passive, consumption-centred view of leisure found in neoclassical economics and instead focused on the creative content of leisure time. The latter, in a workless future, would become the basis for great art and beauty in society. Keynes's concern was that it would take time for humanity to kick the work habit and that the full potential of work reduction would be realised only gradually.

What Keynes failed to contemplate, however, was the scope to elevate the quality of work. He missed how in the harnessing of technology drudgery could be eliminated while realising the value of work. Working less through automation in this respect could be seen as consistent with improving the quality of work. Keynes's commitment to the conventional economic theory of work blinded him to the scope to achieve 'good' work.

Here lessons can be drawn from Marx. In the latter's vision, as mentioned above, the objective of automation was not to eliminate work because of its inherent disutility but instead to negate the alienation of work under capitalism. Marx recognised that for technology to be used in the service of shorter work hours changes in ownership relations would be required. In effect, common ownership was the only way to guarantee a form of automation that delivers shorter work hours. But in the course of changing ownership relations the character of work could also be changed. Working under more cooperative conditions, workers would come to experience the rewards of work and in this sense would be able to realise their humanity in work, rather than just outside of it. Marx's appeal to communism, in essence, rested on the belief that it could deliver a richer form of existence both within and without work (see Sayers 2005; Spencer 2018).

The summary point here is that automation ought to be a means to less and better work. Here emancipatory goals extend to realising the benefits of work, not just leisure. The vision presented is the very opposite of the one presented in economics. Instead of seeking to promote consumption at the expense of less work, there is a broader focus on the need to meet human creative needs in activities that in themselves enrich life, rather than denigrate it. Rejecting economics, the case is made for using technology to transform work and life in ways that add to human well-being.

Conclusion: Reworking Economics

The idea of work has been an important blind-spot in economics. While economists have made reference to the activity of work in their writings, they have failed to address in any detailed way the content and quality of work. How work affects peoples' lives has been systematically ignored in economics.

This neglect reflects deeper biases and flaws in economics. Economists have tended to see humans as consumer hedonists—they have recognised work only to the extent that it offers a means to consumption and denies leisure. There has been no acknowledgement that humans have needs for creative activity and that they might be drawn to work itself as a source of fulfilment.

Modern perspectives linked to the economics of happiness and behavioural economics, to be sure, address the non-pecuniary motives to work. But they do so in ways that leave broader questions unanswered. Work, following the formal method of neoclassical economics, is treated as just another variable in the individual's utility function. The wider significance of work in human life continues to be missed.

The ideas of heterodox economists on work remain insightful. As argued above, there are important ideas to be taken from the writings of Marx and Veblen on work. These ideas show how attitudes to work are malleable and dependent on the organisation of work. They also highlight the scope for humanising work in ways that enhance well-being. It is to the detriment of the economic analysis of work in particular and economics more generally that heterodox economic ideas remain excluded from economics debates.

Finally, as addressed above, there is the issue of automation and the future of work. When seen through the lens of economics, work in an automated future appears much like the present—that is, work remains a disutility. There is no sense of how automation can be used to promote higher quality work alongside and in addition to more free time. There is a lack of vision for creating a better world where technology operates to enhance life both within and beyond work. The conclusion is that if we are to fully understand the possibilities for work and life in the present and in automated futures to come we must overhaul and ultimately transcend the economics of work.

References

Autor, D. (2015). Why are There Still So Many Jobs? The History and Future of Workplace Automation. *Journal of Economic Perspectives, 29*(3), 3–30.

Clark, A., & Oswald, A. (1994). Unhappiness and Unemployment. *Economic Journal, 104*(424), 648–659.

Ford, M. (2015). *The Rise of the Robots. Technology and the Threat of Mass Unemployment*. London: Oneworld.

Gorz, A. (1985). *Paths to Paradise: On the Liberation from Work*. London: Pluto Press.

Jensen, C., & Meckling, W. (1976). Theory of the Firm: Managerial Behavior, Agency Costs, and Capital Structure. *Journal of Financial Economics, 3*, 305–360.

Keynes, J. (1930). Economic Possibilities for Our Grandchildren. In J. Keynes (Ed.), *Essays in Persuasion* (pp. 358–373). London: Norton.

Layton, E. (1962). Veblen and the Engineers. *American Quarterly, 14*(1), 64–72.

Marx, K. (1977). *Economic and Philosophic Manuscripts*. London: Lawrence & Wishart.

Mokyr, J., Vickers, C., & Zierbarth, N. (2015). The History of Technological Anxiety and the Future of Economic Growth: Is This Time Different? *Journal of Economic Perspectives, 29*(3), 31–50.

Sayers, S. (2005). Why Work? Marxism and Human Nature. *Science and Society, 69*(4), 606–616.

Skidelsky, R., & Skidelsky, E. (2012). *How Much is Enough? Money and the Good Life*. London: Penguin.

Smith, A. (1976). In R. H. Campbell & A. S. Skinner (Eds.), *An Inquiry into the Nature and Causes of the Wealth of Nations*. Oxford: Clarendon Press.

Spencer, D. A. (2009). *The Political Economy of Work*. London: Routledge.

Spencer, D. A. (2018). Fear and Hope in an Age of Mass Automation: Debating the Future of Work, New Technology. *Work and Employment, 33*(1), 1–12.

Srnicek, N., & Williams, A. (2015). *Inventing the Future. Postcapitalism and a World Without Work*. London: Verso.

Veblen, T. (1898). The Instinct of Workmanship and the Irksomeness of Labor. *American Journal of Sociology, 4*, 187–201.

Williamson, O. (1985). *The Economic Institutions of Capitalism*. New York: Free Press.

7

Attitudes to Work

Pierre-Michel Menger

My presentation has to do with the attitudes to work in France and Europe, mainly through the channel of the welfarist understanding of work, as it is challenged now by current and increasing job polarisation.

There are two opposed characterisations of work. One highlights its instrumental, monetary value, and the other one highlights its expressive, non-monetary value. The duality of semantics articulates that quite well, opposing labour to work, burden to achievement. It is rather easy to define the negative value of work as this set of painful constraints and efforts that hamper free self-disposal. It is less easy to define the self-fulfilling value of work. It can refer to a parameter within the set of characteristics that attach to jobs, as in the well-known argument of compensating differentials that goes back to Adam Smith (Smith 1776).

A more radical route leads to endorse an ontology that promotes individual achievement and social emancipation through work, rather than through leisure. From the late eighteenth century onwards, the surest

---

P.-M. Menger (✉)
Collège de France, Paris, France
e-mail: pierre-michel.menger@college-de-france.fr

sense of work was found in its productive nature, of which creative labour came to be seen as an epitome (Elster 1985; Taylor 1989; Sennett 2008).

A third layer of meaning came to be added to the concept of work from the late nineteenth century onwards, with the development of the welfare state model and the gradual allocation of social security and rights attached to labour relations. The welfarist doctrine claimed that work is the vehicle to human flourishing, primarily through rising wages and consumption (Habermas 1986).

Can we simply pile up the three layers of instrumental, expressive and welfare related content of work? The interplay of the first two dimensions seems at first sight rather obvious. The instrumental monetary and expressive non-monetary content or functions of work are positively correlated, as summarised by Galbraith. I quote: 'Those who most enjoy work—and this should be emphasised—are all but universally the best paid'. This is accepted. However, this one-dimensional grading has its own limitation. For example, creative workers do not fit that picture well (Garner et al. 2006; Menger 2014). I will not elaborate on that point.

Mainly, I now want to show how the welfarist approach to work attempts to correct or counterbalance the hierarchical evidence of the correlation between instrumental and expressive valuation of work. I will take stock of the French mode—I am French—whether or not there are lessons to be drawn from it to reach a more general understanding of the present and future value of work. This may be disputed.

According to the Organization for Economic Cooperation and Development (OECD) Better Life Index (OECD 2015), the average person has an enviable wellbeing in France (OECD 2017), and yet there is a French mood of everlasting dissatisfaction that generates political swings as well as regular call for structural—that means radical—reforms, especially in the context of the eurozone. There is a French paradox. The French, according to numerous international surveys, are among those who attach the most importance to work and see it as a means of self-fulfilment. At the same time, they are those who wish to, and in fact actually do, devote least time to it, and express strong dissatisfaction with pay and career prospects (Méda and Vendramin 2017). The question might come: are the dissatisfied people building a kind of avant-garde better equipped to face the liberating as well as the threatening dimensions of technological innovations? Let me review four potential explanations for this paradox.

The first signature of the French model is the regulation of the working time and retirement age. The average annual number of hours worked by full-time wage earners in France is the lowest across the European Union. By contrast, the French self-employed workers are true workaholics and among the most zealous in Europe (INSEE 2018). Moreover, the average age at which French workers leave the labour market and retire remains among the lowest in the OECD zone (OECD 2017). Are the French citizens, and French dissatisfied people, spoilt citizens that fear the end of welfare improvement, or is dissatisfaction mainly an issue of composition effects (Murphy and Topel 2016; INSEE 2017)? We should go further in the investigation of the paradox.

The second point is dissatisfaction with pay and low confidence in the future. France's choice has been to reject the 'working poor' model. The minimum wage is among the highest of OECD countries. Over the last 55 years, it has increased faster than the inflation rate and faster than the average wage over the last 20 years. France has indeed a fairly redistributive policy that lowers income inequality and manages to have a rather low share of people below the poverty line. This leads to a wage compression that results from two distinct mechanisms. For the lower part of the wage distribution, the high level of minimum wage dramatically reduced lower tail inequality. For the upper part of the distribution, there is a decrease in the skill premium (Verdugo 2014). Yet, this generates dissatisfaction with pay, especially among those who invest in higher education and expect a good return from it (Artus 2017). The so-called talent drain in France builds on this unbalanced return on education and advancement.

There is now also a growing concern about the momentum that the working poor model gains in France and about the costs of the fairly tight safety net used to buffer it. In fact, the polarisation of the labour market paves the way for a growing structural inequality. Jobs are concentrating at the two extremities: skilled and well-paid jobs in sophisticated sectors, and unskilled and/or deskilled low paid jobs in unsophisticated services. Yet, because low skill, low wage jobs must be created to increase the employment rate, this increase inevitably leads to an increase in income inequality (Artus 2017).

This can be called a curse of higher employment rates. Higher income inequality—yet not the astronomically high rate observed in the US—must be tolerated if the aim is to obtain a higher employment rate.

Can the curse of the high employment rate be averted? It turns out that almost all of the OECD countries that have a high employment rate and low income inequality build their welfare policy on two pillars. Switzerland, Denmark, Sweden, the Netherlands, Austria and Finland have not only large-scale redistributive policies but also a workforce with high labour force skills, including among the low skilled, thanks to a high quality education and vocational training system.

Let us look at the issue of unemployment. The safety net put in place in France has been tightly secure for decades. The high level of employment protection should dampen anxieties, but it is quite the contrary. When asked about how confident they are in their ability to keep their job over the coming months, the French are amongst the most likely to say they are not very confident. For sure, unemployment in France is high and has remained so for more than three decades, yet it mainly hits the low skilled to a greater extent than in the US or the UK, due to the rejection of the working poor model. At the same time, unemployment benefits and unemployment compensation duration in France are among the highest in Europe. The combination of strict employment protection laws and generous unemployment insurance has backed a strong insider/outsider duality in the labour market, with strong discrepancies between permanent jobs and temporary fixed term contract jobs (OECD 2017).

One striking feature of that dualistic structure is the French model of 'flexicurity'. Workers in growing numbers, mainly unskilled and service workers, ultimately compensated unemployment spells with very short-term jobs. As a result, the category of unemployed workers that still work intermittently in order to accumulate unemployment insurance benefits and wages has considerably grown over the last decade. The way most unions operate in France perpetuates this labour market dualism. France has one of the lowest rates of union membership in the OECD, and yet one of the highest rates of wage and collective bargaining coverage, due to the legal and administrative extension procedure, which results in the application of collective agreements to firms that are not members of one of the signatory employer associations. This impacts the way in which unions behave, mainly in a confrontational style. Indeed, we know that better, more cooperative, labour relations positively correlate with union membership rate (Cheuvreux and Darmaillacq 2014).

This leads us to the fourth explanation of the French paradox, which has to do with distrust in labour relations and with managerial flaws. France's overall score of management quality is not that bad, but France's sense of hierarchy and centralisation certainly nurtures a confrontational mood in labour relations (Bloom et al. 2012).

Faced with this paradox, what are the options? Let me consider the usual three suspects: voice, loyalty or exit. Voice means going one step further in reducing the legal work week time, increasing the minimum wage, further reducing income and wealth inequality, and massively investing in public education. That would amount to making France the expected land of higher equality of outcomes, not of opportunity only. That is certainly the option of the French extreme left wing.

Loyalty leads to reform and trust in the improvement potential of the French model. That is certainly the route taken by President Macron now, with its multiple challenges: decentralising labour relations and negotiations, supporting the entrepreneurial spirit, investing in better management, education, lifelong training, building trust and a sense of a social positive-sum game, by securing a higher level of structural growth and building a tangible link between growth, innovation and social mobility. These are major challenges that have been discussed for years, but reform has now gained momentum.

Exit refers to a transformation of the wage-earning society and can take two different ways: the independent work option, or a transformative welfare state that should encompass waged labour as well as independent work. Ideally, the loyalty and exit options should not be exclusive. In the last section of my paper, I mainly focus on the exit option, because it provides a way to extend the discussion of the valuation of work in a context of rising autonomy and independence at work. It may offer us a glimpse into the future of work.

I will leave the two questions that remain in view of independence at work. If self-employment is so desirable, why is the number of self-employed workers not higher (Benz and Frey 2008)? The second one is, how can we explain that a great number of people who enter self-employment and who might do better if salaried persist in independent activity (Rosen 1986; INSEE 2015; Lamarche and Romani 2015)?

Let us go to my conclusion. The premise of an enterprising and ambitious France—that is the present motto—attempts to find its way towards a new, more flexible welfare state. The aim is to pragmatically confront the challenges set at the same time by the labour market polarisation, the digital revolution and the preservation of a European—or we could now better say continental—welfarist model. The challenge is to escape the curse of high employment rates, as well as the pitfalls of the working poor, entrepreneurs or self-employed. Instead of adjusting the existing tools to a rapidly shifting technological and globally competitive environment, one could design a totally new scenario. New? Maybe not that much. Remember that Ronald Coase, a very long time ago, asked why not nexuses of bilateral contract work negotiations instead of firms (Coase 1937)?

What would a flexible welfare state look like? It could be based on so-called social drawing rights (Supiot et al. 2001). The drawing rights framework might build on various existing social rights: assistance for the unemployed in creating or taking over businesses, training leave, training vouchers, special leave, time save accounts, universal basic income, in order to extend them and, more importantly, better manage their allocation, combination and interaction. We should note that this may, in the long term, lead to a management of preferences, rights and risks that would erase the barrier between market, firm and public or private regulations (Menger 2002).

Ironically, based on personal accounts one would draw on, management of one's life course would encompass more and more dimensions: paid work, community work, leave for job search and occupation switch, lifelong training spells. It would, at the same time, be subject to bargaining processes that resemble the running of an individual micro firm, with investment in skill acquisition, portfolio of competences, management of rights, interim devices, arbitrations and so on. Platforms and digital devices should help. After all, the future is made out of tensions to create and then to reduce.

References

Anand, P., Durand, M., & Heckman, J. (2011). Editorial: The Measurement of Progress—Some Achievements and Challenges. *Journal of the Royal Statistical Society, 174*, 851–855.
Artus, P. (2017). *Flash Economics* (daily publication). Paris: Natixis.
Benz, M., & Frey, B. (2008). Being Independent is a Great Thing: Subjective Evaluations of Self-employment and Hierarchy. *Economica, 75*, 362–383.
Bloom, N., Sadun, R., & Van Reenen, J. (2012). The Organization of Firms Across Countries. *Quarterly Journal of Economics, 127*(4), 1663–1705.
Cheuvreux, M. & Darmaillacq, C. (2014). Unionisation in France: paradoxes, challenges and outlook. *TRESOR-ECONOMICS 129*, 1-12.
Coase, R. (1937). The Nature of the Firm. *Economica, 4*(16), 386–405. Blackwell Publishing.
Coe-Rexecode, 2016. *La durée effective annuelle du travail en France et en Europe. Document de travail n°59*, Paris : Coe-Rexecode.
Elster, J. (1985). *Making Sense of Marx*. Cambridge, UK: Cambridge University Press.
Galbraith, J. K. (2004). *The Economics of Innocent Fraud: Truth For Our Time*. Boston, MA: Houghton Mifflin.
Garner, H., Méda, D., & Senik, C. (2006). La place du travail dans les identités. *Economie et Statistique, 393–394*, 21–40.
Habermas, J. (1986). The New Obscurity: The Crisis of the Welfare State and the Exhaustion of Utopian Energies. *Philosophy and Social Criticism 11*(2), 1-18.
INSEE. (2015). *Emploi et revenu des indépendants*. Paris: INSEE.
INSEE. (2017). *Emploi, chômage, revenus du travail*. Paris : INSEE.
INSEE. (2018). *Tableaux de l'économie française*. Paris : INSEE.
Lamarche, P., & Romani, M. (2015). Le patrimoine des indépendants. In *Emploi et revenu des indépendants*. Paris: INSEE.
Méda, D., & Vendramin, P. (2017). *Reinventing Work in Europe. Value, Generations and Labour*. London: Palgrave Macmillan.
Menger, P.-M. (2002). *Portrait de l'artiste en travailleur. Métamorphoses du capitalisme*. Paris: Seuil.
Menger, P.-M. (2014). *The Economics of Creativity*. Cambridge, MA: Harvard University Press.
Murphy, K., & Topel, R. (2016). Human Capital Investment, Inequality, and Economic Growth. *Journal of Labor Economics, 34*(2, Pt. 2), S99–S127.

OECD. (2015). *How's Life? 2015: Measuring Well-being.* Paris: OECD Publishing.
OECD. (2017). *OECD Economic Surveys: France 2017.* Paris: OECD Publishing.
Rosen, S. (1986). The Theory of Equalizing Differences. In O. Ashenfelter & R. Layard (Eds.), *Handbook of Labor Economics.* Amsterdam: North-Holland.
Sennett, R. (2008). *The Craftsman.* New Haven, CT: Yale University Press.
Smith, A. (1776). *An Inquiry into the Nature and Causes of the Wealth of Nations.* London: Strahan.
Supiot, A., et al. (2001). *Beyond Employment. Changes in Work and the Future of Labour Law in Europe.* Oxford: Oxford University Press.
Taylor, C. (1989). *The Sources of the Self.* Cambridge, MA: Harvard University Press.
Verdugo, G. (2014). The Great Compression of the French Wage Structure, 1969–2008. *Labour Economics, 28*, 131–144.

8

Work as an Obligation

Nan Craig

As a concept, work is difficult to define, even though most of us feel that we know which of our daily activities is work and which is not. We disagree and contradict ourselves over the most basic characterisations of work—is it a cost, or a pleasure? Purely a matter of subsistence, or something that imbues our lives with additional meaning and purpose? What would we be without it—free, or lost?

Our attitudes to work are confused in part by being inherited from wildly contradictory traditions—ancient Greece, Puritanism, neoclassical economics. These influence the kinds of work we value, or whether we view a particular activity as 'work' at all. For instance, the late Apple CEO Steve Jobs' exhortation to 'Do what you love, and love what you do' is a prime example. In an essay for the online magazine *Jacobin*, Miya Tokumitsu pointed to this mantra as a pernicious aspiration that manages to both exalt and undermine work, implying that the only really valuable labour is the enjoyable and satisfying (and photogenic)

N. Craig (✉)
Centre for Global Studies, London, UK

kind.¹ By implication, if you do happen to love your work, you should want to do it all the time. You shouldn't, of course, be so mercenary as to expect to be paid for it.

In one sense the slogan of 'Do what you love' reflects the split between 'labour' and 'work' that Hannah Arendt identified, tracing the etymologies back to Greek conceptions of work and life.² 'Labour', she said, is the endless, necessary work required to keep human life running smoothly: growing and preparing food; cleaning and fixing; bearing children and raising them. 'Labour', while it can be done out of love, is not *only* done out of love. It's also done because life itself will fail if it is left undone.

'Work', on the other hand, is the act of creating something permanent, outside of ourselves. The Greeks saw this as a higher calling; creative 'work' set humans apart from other creatures. Finally, what Arendt called 'action'—in Greek, *praxis*—was for the Greeks the highest pursuit, that of debating and decision-making with other citizens. Slaves and women needed to labour and artisans needed to work so that citizens could be free from the daily grind and participate in a more fulfilling political life.

Arendt herself, while not as dismissive as the Greeks were of labour, argued that in a consumer society, everyone becomes a labourer rather than a worker, since everyone is only working in order to keep consuming. Doing work for its own sake makes you a worker, rather than a labourer. 'Do What You Love', then, performs a similar sleight of hand: by ignoring the constant, unavoidable labour that goes into maintaining life, and concentrating only on those individuals lucky enough to be working out of enthusiasm, it manages to make most of the world of work disappear.

The Stoic philosophers were the first Western thinkers to break with the Greek ideal and argue that hard work could be valuable and noble in itself. Their influence on the early Christian church helped embed that ethos—though for hundreds of years, work still only came second in importance to religious devotion. Catholic monasteries were hives of activity but also, as Herbert Applebaum points out,³ surprisingly enthu-

---

[1] Tokumitsu (2015).
[2] Arendt (1958).
[3] Applebaum (1992).

siastic adopters of automation technology, since the principle that it was good to toil but *better* to pray meant that reducing labour time was not unwelcome. The great shift in Christianity's attitude to work came with the Reformation and the growth of Calvinism and Puritanism, both of which placed hard work as the highest form of service to God. This wasn't work as joyful but rather work as penance—and Puritans, in particular, believed that work should be undertaken to maximise profits, which then ought to be spent on charity and good works. Max Weber argued that this (literal) faith in profit-making through maximum efficiency and constant toil was the fertile ground that allowed capitalism to develop. This belief in work as character-forming (and failure to work at maximum effort as wasteful and morally suspect) lingers in the benefits system, in the idea of forcing people to do pointless make-work as a condition of receiving subsistence.

As the industrial revolution moved workers out of the home and off common land and into factories, farms and offices, work became more solidly defined as waged labour. As Andrea Komlosy describes in her chapter, labour outside of 'workplaces', in the home or elsewhere, stopped being considered work at all. Alongside the technological revolution, the development of economics meant the growth of a conception of work that focused on its costs, rather than work as fulfilling.[4] Since the early twentieth century, labour organisations, governments and employers have defined work as paid employment, or self-employment. The broad idea of *work* turned into narrowly defined *jobs*.

But this definition constantly butts up against the colloquial, much broader, idea of work. Parenting is often claimed as hard work, for instance. *Art is work*, argue artists—often as they try to negotiate decent pay for their creations. Care work—both that of family members and of unrelated paid and unpaid carers—has been identified by feminist economists as a hugely undervalued contribution to the economy. This is in part the result of that pervasive, Puritan ethos that those who don't work are lazy and immoral; we need to justify what we do by arguing for its status as 'real' work. The question of 'who works' is a political one. But it

---

[4] Spencer (2009).

also reflects a sense that these endeavours *are* hard work, with valuable results. It can seem difficult to establish a clear thread connecting these diverse human activities. Once we accept that work is bigger than the idea of the paid job, it begins to spill into every area of life.

So what makes work *work?* There is, in fact, a common element that unites wage-labour, service to others, vocation, parenting, art, subsistence, and the various other kinds of human exertion that we bundle together under the label of 'work'.

That common thread is a sense of obligation—or, more positively, of responsibility. Work is the thing that you get up and do, *whether you want to or not*. Sometimes the obligation or responsibility is primarily to oneself or one's family—you keep going back every day to the miserable job because it keeps you or the people you love in food and shelter. But most people also have at least a nominal sense of responsibility to their employer, or to their co-workers or clients, not to leave them in the lurch. At other times the feeling of work comes from a perceived responsibility to the needs of another person—as when we take care of children or someone who's ill. Sometimes the impulse to work may be felt as an obligation to develop one's talent. And artists often feel an obligation to the work itself, to bring it into the world. Work happens wherever we feel something is demanded of us—whether that's the urge to clean up a messy kitchen or the urge to play a piece of music beautifully.

Its connection with responsibility is also the reason work is so intertwined with meaning. One of the odd things about work which the 'do what you love' mantra overlooks is that people often don't know what they love to work at until they've been doing it for a while. Many people fall into a line of work that eventually begins to intrigue and fulfil them, and sometimes even unpleasant work can be made more meaningful by a commitment to doing it well. It is pointless, though, to insist that people should be able to find meaning arbitrarily in their work, whether it's useful or not. Depressingly, as David Graeber argues in *Bullshit Jobs*, a lot of work doesn't really demand to be done at all.[5] If we agree that a sense of responsibility—a sense that the work demands that you do it—is a defining feature of work, then a bullshit job isn't properly work at all.

---

[5] Graeber (2018).

In this sense, work is clearly relational; it implies a relationship *to* or *with* something, whether that something is another person, an organisation, or reflexively with oneself. The obligation may even be towards non-humans. Veterinarians, for instance aren't generally in the job because they feel a responsibility to pet *owners*. And implicit in much environmental activism is the idea that we owe responsibility to the planet itself, not just as our home but as another living thing.

The idea of obligation or responsibility doesn't, of course, preclude resenting the work you do, or even hating it. The obligation itself might be experienced narrowly as a contractual one, or more deeply as a moral one—or even as a social or peer pressure. There are plenty of unhealthy elements to the work-lives of social media entrepreneurs making names for themselves on Instagram or YouTube, but one element that seems to weigh heavily is the sense of obligation to fans. People doing that work report feeling under pressure to turn every element of their life into a show to please followers and brands. A heavier sense of demand doesn't necessarily lead to more meaningfulness—it can simply lead to a terror of letting people down. It seems likely, though, that the stronger the *moral* obligations one feels are connected to one's work, the more meaningful it would be.

I am not arguing that obligation is the same as work—clearly there are many forms of legal and moral obligation that are not work. But I'd argue that the work relationship always has some element of obligation.

One other result of thinking of work as a relationship of obligation is that it doesn't force us to answer the question of whether work is a cost or a pleasure; as an obligation it may be either or both. All obligations carry a certain onerousness simply because you can't stop whenever you like. In this way, we can draw a hypothetical line between work and pure leisure, even when both are unpaid: a leisure activity is the one you feel able to pick up and put down with complete freedom. That line must remain fuzzy, I think. But if you feel very guilty for not practising at piano or finishing the crossword, then your hobby has probably become at least slightly tinged with the character of work rather than play.

What does this mean for questions about the future of work? There are broadly two perspectives on the importance of waged work (specifically of *jobs*) for humans. The first rests on the belief that work, even sometimes quite low-quality work, is essential to human flourishing by providing structure, identity and self-respect. The opposing view is that full-time paid work is an encumbrance forced on us by necessity (or by an unfair economic system), and that most people would find their lives more fulfilling and richer if they could reduce paid work as much as possible.

This division crosses the political spectrum, but in the debate around automation it is often expressed as either enthusiasm or dismay at the idea of a Universal Basic Income. A Universal Basic Income (UBI) is a non-means-tested payment made to everyone, independent of employment status or qualification for other social benefits. People who think that a UBI would be disastrous tend to reference the effects of long-term or mass unemployment, and the sense of hopelessness and inactivity induced by sudden compulsory redundancy. This reflects Marx's vision of work as essential to human flourishing. Enthusiasts for UBI, on the other hand, tend to either take the economists' view of work as a cost or to come from a left or feminist tradition which is suspicious of the idea that paid work is a choice at all.

The idea of work as obligation, however, seems to support the idea that a world without paid work would not be a world without responsibility—and by extension, the structure and meaning that responsibility gives to life.

Fear of sudden mass unemployment is reasonable, but fear of reduced work is not. Sudden loss of structure and daily purpose in life would be a dangerous prospect for most people, but there are other models of work reduction to consider. Research on retirement, for instance suggests that people who have thought ahead and planned how they want to live manage better and have healthier retirements than those who haven't thought it through (assuming similar resources). Some people crave external structure to their days, while others feel comfortable creating their own. Admitting that individuals differ in this sense only means that we need to create alternative structures, not scrap the idea of freedom from jobs entirely. It is possible to reconcile the idea that humans need purposeful work (and sometimes even structured obligations) with the idea that *fewer jobs* could be desirable, as long as we acknowledge that life is full of unpaid obligations.

If we pull back from distant-future utopias and address the here-and-now instead, anti-work arguments are a reasonable corrective to our excessive valorisation of work. Feminist anti-work arguments (for instance in Kathi Week's *The Problem with Work*[6]) are particularly strong, contradicting the liberal-feminist and Marxist-feminist assumption that full-time paid work is unequivocally a good thing for women. The full entry of women into the job market has been achieved with no systematic provision at all for reassigning all the non-paid work they had previously done. While there is much talk about 'work-life balance', for most women with children it's more about 'work-work balance'—juggling an endless stream of obligations with little to no leisure.

If we agree that invoking a sense of responsibility—of being needed—is a defining effect of work, and that it can help us build a stronger sense of meaning in life, then it helps explain why many people feel *any* work is preferable to being completely unoccupied. Also, why jobs which are completely pointless are so demoralising. The sociologist Henrik de Man believed that humans needed and naturally tended to enjoy work (under the right circumstances) and found inactivity torturous.[7] Backing this up, research on basic income pilot schemes has found that people receiving guaranteed income tend only to reduce their paid work to the extent that they have other demands on their time—in particular people with young children, or students.

A bigger question is whether increased automation can deliver any of this. Within the current economic system, the answer is: probably not. Large-scale redistribution, or at least ensuring that profits from productivity went to workers rather than share-holders, would be a minimum requirement. In order to make it feasible for people to spend less time in paid work, they need to be paid more. If automation does require us to work less, we shouldn't fear it. More important is that we plan for that possibility. Where paid work disappears suddenly, it can leave a vacuum. We need to make sure that new responsibilities and obligations have time to grow into the space where paid work used to be.

---

[6] Weeks (2011).
[7] De Man, quoted in Applebaum (1992).

References

Applebaum, H. (1992). *The Concept of Work: Ancient, Medieval and Modern*. Albany: State University Press of New York.
Arendt, H. (1958). *The Human Condition*. Chicago: University of Chicago Press.
Graeber, D. (2018). *Bullshit Jobs: A Theory*. London: Allen Lane.
Spencer, D. (2009). *The Political Economy of Work*. London: Routledge.
Tokumitsu, M. (2015). Forced to Love the Grind. *Jacobinmag.com*. [Online]. Retrieved February 4, 2019, from https://www.jacobinmag.com/2015/08/do-what-you-love-miya-tokumitsu-work-creative-passion/
Weeks, K. (2011). *The Problem with Work*. Durham, NC and London: Duke University Press.

Part III

Attitudes to Technology

9

Attitudes to Technology: Part 1

James Bessen

I am going to lay out three economics propositions about the impact of AI in the near term, regarding automation in particular, the near term being 10 or 20 years. First, most automation has been partial automation rather than complete automation. Second, partial automation can lead to increases in employment in affected industries as well as decreases. Third, even if automation does not destroy jobs on the net, it will still be highly disruptive because people need to learn new jobs and skills in order to remain employed.

First, it is important to distinguish between automating a task and automating a job. Jobs involve many different tasks, often very diverse skills. Because of that, it is relatively rare that a job will be completely automated.

For instance, I looked at the number of detailed occupations listed in the 1950 US census and tracked how they are categorised today. Which of them have disappeared because of automation or other reasons? First off, most of them still exist in one form or another today, maybe with a

J. Bessen (✉)
Boston University School of Law, Boston, MA, USA

title change. A number disappeared because demand fell, for a variety of reasons: changing tastes, changing patterns. For instance, the occupation of 'boarding house keeper' is no longer listed. Some disappeared because of technological obsolescence, when the industry was replaced. There is no longer 'telegraph operators' listed as a category.

Only one occupation was completely automated, and that was the job of elevator operator.

Now we are seeing machine learning, where we have all these capabilities, where machines can do better than humans on certain tasks. They can read lips or read x-rays better than humans. Machine learning systems tend to be exceptionally good at individual tasks, typically tasks where there are large amounts of labelled data. They do not have common sense or meaning in the sense that we attach meaning to things. Machine learning has been around since the 1980s and has been enhanced by improvements in software, but also because we have much greater access to large pools of data and much bigger hardware to throw at it. Nevertheless, the improvements made have been on specific tasks.

Consider how automation has affected three occupations. The first is accountants.

Accountants' tasks were first automated by computers in 1954 and there have been large improvements since then. We are surveying artificial intelligence developers, getting a sense of what is being automated today. These new systems can do sophisticated things like read through large databases of contracts and pull out relevant contract terms for an auditor. But there are many tasks that these systems cannot do, such as to review and advise on organisational controls.

The first computer automation for loan officers occurred in 1956 and the first AI systems were put into commercial use in 1987 for automatic credit card detection. Automatic loan processing for some kinds of loans has been in place since 1972. There have been enormous improvements since then, but again there are many tasks that no one is automating today such evaluating business performance of commercial lendees.

The first systems to automate tasks for paralegals and legal assistants were built in 1963. Since the late 1990s there has been a boom in applications for electronic text retrieval. These systems, both AI and algorithmic, are used to identify documents relevant to litigation and, by some measures, they can do the job better than humans. But they cannot do the whole job. Because of that employment of paralegals in the US, where these systems are most widely used, has actually grown over the last 15 years.

These cases suggest that over the next 10 or 20 years, AI is going to have a big impact, but it is mostly going to be in the sense of partial automation. There are going to be some important exceptions: certain jobs for truck drivers and warehouse workers are going to be completely automated.

Given that most of the automation is partial, we need to recognise that automation can create jobs and, in some cases, will. Even in the affected industries, jobs can increase. Look, for example, at the US textile industry (Bessen 2015, 2019). We are accustomed to associating automation with terrible job losses in industries such as textiles and that has been the experience of the last several decades, but it was not the case earlier. By 1910, the textile industry had already become highly automated with up to 24 looms used per weaver at the best mills. Effectively, over the course of the nineteenth century, 98% of the labour that went into weaving a yard of cloth had been taken over by machines. Nevertheless, if you look at the employment of production workers in cotton textiles, the nineteenth century and well into the twentieth century saw a high rate of automation accompanied by rapid job growth.

This is an interesting puzzle. What changed here? Demand changed. Because of automation, less labour was required to produce a yard of cloth, which meant that, in a competitive market, firms charged lower prices. Because they charged lower prices, many more yards of cloth were consumed; people bought more at lower prices. If we look at the per capita consumption of cotton and synthetic cloth in the US, it went up 20-fold, over 10-fold during the nineteenth century alone. This made a huge difference. In the early nineteenth century people had very little cloth. It was very expensive. The typical person had one set of clothing.

When the price went down, there was a large pent up demand that meant that people could afford more cloth and they bought a lot more cloth. They bought so much more cloth that even though it required less labour time to produce a yard of cloth, so many more yards were bought that the employment went up. But when you come to the mid-twentieth century, we have full closets. We have cloth being used for upholstery, all sorts of uses, and demand is satiated. Automation is still increasing labour productivity at about 3% a year, as it had been for the last 200 years, but this no longer generates increased demand. What happens? The labour-saving effect dominates.

We see similar things today. Barcode scanners came in and reduced the amount of time required for a checkout cashier, but we have more cashiers. E-commerce has actually led to increased jobs. Electronic discovery, which I talked about, has led to an increase in paralegals. My favourite example is the automated teller machine and bank tellers (Bessen 2016). Between 1995 and 2005, in the US we had a huge influx of automated teller machines, but the number of bank tellers went up. It is a similar demand story. It meant that banks could operate branches with fewer tellers, so it was cheaper, and so, because they could reach unmet demand in terms of regional areas where people wanted cash, they could open up many more branches, so many more branches that the number of tellers increased.

This pattern is general through much of the non-manufacturing sector, where the effect of information technology has been to increase job growth rather than decrease it. In manufacturing, it has been the reverse. We have had 200 years of automation in manufacturing and in many markets, demand is satiated. In the near term, demand elasticity is not going to change very much. The effect of AI in the near term is going to be very similar to the effect of the computer automation we have had over the last 50 or 60 years. We are going to see some occupations grow. We are going to see some decline. There are big questions about how long that pattern will last, and longer term poses a different issue. I will get back to that briefly.

The third point I want to raise is it is not all a happy story. This is a point Carl raised earlier that automation raises inequality in multiple dimensions. Consider the job of the typesetter (Bessen 2015). This job has largely disappeared thanks to desktop publishing, but there has been a substitution where desktop publishers and graphic designers are now producing the text we read. The number of graphic design jobs has gone up much more than the loss in typesetter jobs. I think this is the real disruption. People talk about and think about the disruption from AI being one of machines replacing humans. It is really a story about humans using machines replacing other humans.

That ties into all sorts of age-old problems we have had in terms of inequality. I will highlight a few of the ways this plays out. We have new jobs. The graphic designers and desktop publishers need new skills. These skills change rapidly. With new technologies, they are often not standardised. This means that schools have a hard time keeping up. This means that labour markets do not necessarily recognise the skills and pay for them, reward them appropriately. The result tends to stagnant average wages, but also growing inequality, because some people can survive in that environment and prosper, while others cannot. It means different transitions, that people are often in the wrong industries, the wrong jobs, the wrong locations even, requiring a move in multiple dimensions.

It also means greater inequality among firms. Some firms are very good at hiring and training and developing new skills, and others are not. Across the developed world, we are seeing a growing gap between the most productive firms in each industry and the average firm. This is actually fairly dramatic and it is directly tied to technology. In some of my research we find that across industries the more IT they are using, the greater their productivity per worker, but that is especially true for the top four firms in those industries. There is a growing gap between the top four and the rest. This leads to rising concentration across all sectors: manufacturing, finance, service, retail, wholesale. In each industry we are seeing the top four firms dominating the market much more.

Those are my three propositions. I think we have some evidence that these are trends to some extent. We can look back over the last 10, 20, 30 years and see these trends are in place. My basic notion is that, to some extent, these trends are going to continue for the next 10 or 20 years. Longer term though, we are going to face some different problems. One is that, as the technology gets better, we are going to see more occupations that are completely replaced. Two, if we think about the rise and fall of textile workers, that pattern of rising employment followed by sharp declines was mirrored in all sorts of manufacturing industries and sectors. That includes not only developed countries, but in developing nations today. More markets are going to become satiated. Banking, finance and healthcare may be automated today, and they may be in a stage where there is a lot of unmet demand, so there may be job growth. In 30 or 50 years from now, that may no longer be the case. Perhaps we will meet all of our financial needs, in some sense. Then of course there is this question about general AI coming along.

The general takeaway is that demand matters. Repeatedly over the last 200 years we have had various people concerned about the effect of automation on jobs. These predictions have typically not been borne out. That is not a reason to say that predictions today are necessarily wrong. I think the important thing I want to say here is that the reasons they have not been borne out is because people have underestimated the depth of human demand, new demand that is generated. In a general sense, jobs will persist as long as there are major sources of unmet demand that can be satisfied through the market place.

Those are two important conditions, but this really raises and gets to some of the broader issues we just started talking about. What is it that humans want? Are there markets or industries where demand will not be satiated, where technology will never meet all of our needs and wants? Will we ever run out of a desire to have better health and longer life? Is an important part of our consumption really human interaction, or is this something where robots can frighteningly replace or substitute for humans in areas where you might think social interactions were important? We also have what economists call status goods, for example fashion. We demand things because they set us apart from everybody else. Maybe there is always a relative rather than an absolute value to these products. Anyhow, those are some comments and things to think about.

References

Bessen, J. (2015). *Learning by Doing: The Real Connection Between Innovation, Wages, and Wealth*. Yale University Press.

Bessen, J. E. (2016). How Computer Automation Affects Occupations: Technology, Jobs, and Skills. Boston Univ. school of Law, Law and Economics Research Paper, 15–49

Bessen, J. E. (2019 forthcoming). Automation and Jobs: When Technology Boosts Employment. *Economic Policy*.

10

Attitudes to Technology: Part 2

Carl Benedikt Frey

I'm going to spend the next 15 minutes of my talk arguing that we shouldn't feel all that reassured if the future of automation mirrors the past. True, the dystopian belief that technology will make human labour obsolete, prompting mass unemployment, has been proven wrong over the past 200 years. But there have been episodes when things didn't work out well for labour. Over the past 30 years, real hourly wages among prime aged men in the US with no more than a high school degree have been falling. Likewise, during the first 70 years of industrialization, between 1770 and 1840, weekly wages were stagnant in Britain. Taking into account that working hours expanded by some 20 percent over this period, it seems that real hourly wages even fell for a large share of the population.

The point I would like to make is that the price of progress paid by the workforce has varied greatly throughout history, shaping attitudes towards technological progress in the process. New technologies may

---

C. B. Frey (✉)
Oxford Martin School, University of Oxford, Oxford, UK
e-mail: carl.frey@oxfordmartin.ox.ac.uk

either augment our skills in a happy way, allowing people to earn better wages. Or it may replace us in exiting jobs and tasks, reducing the earnings potential of groups in the labour market, whose skills become obsolete. This is what has happened to large swathes of the population since the Computer Revolution took off in the 1980s. As middle-income manufacturing jobs have dried up due to globalization and automation, the earnings capacity of men who didn't go to college has steadily diminished.

The potential set of tasks that computer technologies can perform has expanded over the years. If we go back to the early part of the twentieth century, humans were computers. A computer was an occupation doing basic arithmetic and tabulating the results. Any distinction between humans and computers was therefore meaningless. But after the invention of the first electronic computer, the Electronic Numerical Integrator and Computer (ENIAC), at the University of Pennsylvania in 1946, automation began to spread from one task to another, though computerization wasn't sufficiently cheap to have any meaningful impact on labour markets until after 1980. Yet up until very recently, automation was still confined to rule-based activities that a programmer could specify in computer code.

Recent advances in machine learning is what has brought us all here today. So presumably I do not have to convince you that automation is becoming increasingly pervasive. In essence, the main difference to rule-based age of computing is that top down programming is no longer required for automation to happen. Instead of having a programmer specifying what a computer technology must do at any given contingency, computers can now infer the rules themselves through "examples" or "experience" provided in what is known as "big data". As is well-known, to beat the world champion at Go, AlphaGo drew upon a training dataset of 30 million board positions from 160,000 professional players. Thus, its experience was far greater than that of any professional Go player.

This way, computers are already learning how to perform a variety of non-rule-based tasks, like diagnosing disease, writing shorter news stories, and driving trucks, which were non-automatable only a decade ago. So back in 2013, together with a group of researchers at our engineering sciences department in Oxford, Michael Osborne and I set out to determine the potential scope of automation in the age of machine learning.[1] Because the recent inroads of automation are many, we began by asking the question: in which domains do automation technologies still perform poorly despite recent advances in machine learning? Broadly speaking, we found that humans still hold the competitive advantage in three broad domains: creativity, complex social interactions, and the perception and manipulation of irregular objects. To take one example, the state-of-the-art of technology in reproducing human social interactions is best described by the Loebner Prize—a Turing test competition—where chatbots try to convince human judges that they are actually chatting with a person.

Some pundits have argued that there was a breakthrough in 2014, when one chatbot actually managed to convince 30 percent of judges of it being a human. But it did so by pretending to be a 13-year-old boy speaking English as his second language. And if you think about the variety of much more complex in-person interactions many of us do in our daily jobs, like trying to persuade people, assisting and taking care of customers, managing teams, and so on, algorithms are nowhere near being capable of replacing us in those tasks.

The same is true of many seemingly simple tasks, such as the perception and manipulation of irregular objects. It is quite extraordinary to think that more than 20 years ago IBM's Deep Blue beat Kasparov at chess. Yet even today, there is no robot that is capable of picking up individual chess pieces to clean them and putting them back on the board. In the same way, we struggle to automate a lot of things that require interaction with unstructured environments and irregular objects. For example one of the last things we are likely to automate is the jobs of janitors.

In part, we can circumvent some of these bottlenecks by task simplification. For example we didn't automate the work of laundresses by building robots capable of carrying wood and water into the home yet carrying out the motions that handwashing entails. We did so by inventing the electric washing machine, which does an entirely different set of tasks, but still accomplishes the same goal: clean clothing. And similarly, we didn't automate away the jobs of lamplighters by building robots capable of climbing lamppost. One reason why many commentators underestimate

---

[1] Frey and Osborne (2017).

the potential scope of automation is that they assume that the technology must be capable of performing the exact tasks same that a human does. Yet that is rarely the case. Simplification is mostly how automation happens.

As some of you will know, we reached the conclusion that roughly 47 percent of American jobs are exposed to automation in our 2013 paper. Our finding has often been taken to suggest that all of these jobs are going to disappear in a couple of decades, leaving many unemployed. But we suggested nothing of the sort. What we found is that about half of current jobs are automatable from a technological capabilities point of view, given state-of-the-art computer-controlled equipment and conditional upon the availability of relevant big data.

Now, it has been argued that we shouldn't worry too much about automation because individual tasks are likely to be automated rather than entire jobs. However, many jobs, like those of elevator operators, lamplighters, switchboard operators, farm labourers, and car washers, just to name a few, have been fully automated in the past. What's more, several studies have shown that jobs which are intensive in automatable tasks have experienced employment declines in recent decades. But of course, it is true that many jobs have changed rather than disappeared. That, however, might be of little assurance from the viewpoint of the individual being affected, as it often means that an entirely different breed of workers is required.

If we take the frequently used example of the bank teller, it is true that there are still people working as tellers, despite the proliferation of ATMs. But the skillset of the bank teller of the 1970s has already been rendered redundant. A bank teller who can only accept deposits and pay out withdrawals won't get a job at a modern bank. A teller today is more of a relationship manager at the bank branch. To take another example, if you were to introduce a farm labourer from 1900 on a modern farm, he or she would for the first time encounter tractors, automobiles, milking machines, electricity, computers, GPS technology, and so on. It would take many months if not years to train that person. On paper it is still the same occupation, but in actual fact it's a completely different job.

It must be noted, however, that even if entire occupations vanish or change so much that they remain only recognizable on paper, the implication is not the end of work. One reason is that new jobs and tasks appear as automation progresses. Few of today's jobs existed in 1750 at the dawn of the Industrial Revolution. And job titles like robot engineer, database administrator, and computer support specialists, didn't even exist in occupational classification as late as the 1970s. And while today's tech industries don't employ as many people directly as the smokestack industries of the twentieth century did, they support the incomes of many low-skilled service jobs, as high-income earners, whose skills are complemented by computers, demand many services, like those of janitors, gardeners, hairdressers, fitness trainers, and so on.

Second, even if many jobs are becoming automatable, that doesn't necessarily mean that they must be automated anytime soon. Decisions to automate depend on many factors beyond technology itself. Regulation, consumer preferences, and the relative costs between capital and labour, among many other variables, shape decisions to automate. Technology adoption is never frictionless, and the transition will therefore be gradual. Indeed, it is important to remember that as late as 1900, 40 percent of American's worked on farms. Today, that figure is below 2 percent. Seen through the lens of history, the potential scope of automation today is probably not unprecedented.

Third, technology is not an unstoppable force. Especially when it takes the form of capital that replaces workers. In the end, technological progress depends on societal acceptance for it. And when it threatens people's jobs and incomes, favourable attitudes towards technological change cannot be taken for granted.

When I first presented our paper back in 2013, one economist in the audience dismissively remarked, "Is this not just like the Industrial Revolution in England?… Didn't machines displace jobs then as well?" His comment was meant to suggest that will robots make us wealthier and allow people to take on better paid jobs as technology has in the past. And this has unquestionably been the long-run story over the past 200 years. But the short-run can be a very different matter. And what economists regard as the short-run can be a lifetime for some.

If we go back to the early days of industrialization, when the mechanized factory displaced the domestic system, we know that middle-income artisan jobs dried up, the labour share of income fell, profits surged, and income disparities grew. To be sure, new jobs were created in the factories as artisan jobs disappeared, but many of them were taken on by children. Indeed, textile machines of the early years of industrialization were specifically designed to be tended by children. They were the robots of the Industrial Revolution. Besides working for very little, they were easy to enforce the factory discipline upon. Augmented by machinery, they replaced adult craftsmen earning decent wages. And as their incomes disappeared, many had to reduce their standard of living.

In this light, economists and economic historians have puzzled why artisan workers would have accepted the Industrial Revolution if it reduced their utility. But this is only a puzzle in the absence of cohesion, and cohesion was far from absent. Angry workers rebelled against the mechanized factory on many occasions. And how did the British government respond? By deploying troops against the rioters. The 12,000-man army sent out against the Luddites was even larger than the one Wellington took against Napoleon in 1808.

The age of automation, which took off with the Computer Revolution of the 1980s, in many ways mirrors the experience of the British Industrial Revolution in economic terms. Since the dawn of the computer era, middle-income jobs have disappeared, wages have been stagnant, and the labour share of income has been falling. And there is now quite a significant literature on trade and technology shocks and their impacts on the communities in which people live. Economists have found that places that have been exposed to Chinese import competition and robotization have suffered job losses. And where manufacturing jobs have disappeared, public services have deteriorated, healthcare outcomes have worsened, crime rates have gone up, and marriage rates have gone down. People living in those places have not seen the gains from trade and technology. In fact, they have arguably even been made worse off by it.

So why has automation not been accompanied by similar opposition to new technology as was the case during the Industrial Revolution? In some ways, it already has. To be clear, I am certainly not suggesting that automation is the only reason for Donald Trump's appeal. But it is hard to believe that he would have won the 2016 election if there were well-paying jobs in abundance for the unskilled and wages were rising for everyone. Those who voted for Trump are in essence those who have seen a reversal of fortunes over recent decades. Before the dawn of the Computer Revolution of the 1980s, people could find decent-paying jobs in the manufacturing industry without going to college. As those jobs disappeared, unskilled men who once faced abundant opportunity in the manufacturing industry have felt the force of automation most keenly.

The adoption of robots is highly concentrated in three states that had voted for the Democratic presidential candidate in every election since 1992: Wisconsin, Michigan, and Pennsylvania. And those were the states that swung the election in favour of Trump. While counterfactuals must always be taken with a pinch of salt, my own work with Thor Berger and Chinchih Chen suggest that if robotization had not increased since the last election, those states would have swung in favour of Hillary Clinton, leaving her with a majority in the electoral college.[2]

Attitudes towards automation itself are also already changing. A 2017 Pew Research survey suggests that 58 percent of Americans think that there should be limits on the number of jobs businesses can replace with machines. And 85 percent think that automation should be restricted to hazardous jobs. And unsurprisingly, the unskilled are much more likely to favour restriction on automation.[3]

As I argue in my book, The Technology Trap, the point is that the pervasiveness of automation and people's reactions to it are likely to jointly determine the future pace of automation.[4] Most economists and pundits would probably agree that it makes little sense to think about the future of globalization without considering its political economy: the Trump administration's trade war with China has clearly changed the rules of the game. In a similar fashion, if people don't see the gains from automation trickle down, robots could become the political target that globalization already has. Many Americans already favour restrictions on automation. And populists may eventually supply such policies.

---

[2] Frey et al. (2018).
[3] Gramlich (2017).
[4] Frey (2019).

Current economic trends may of course not continue indefinitely. But there are good reasons to think that the unskilled will continue to see their earnings potential diminish over the next decades. According to our estimates, low-skilled, low-income jobs are most exposed to recent advances in machine learning. Even if the next generation goes to college and successfully acquires new skills, that will be of little reassurance to the generation that's already in the labour market today.

During the British Industrial Revolution, it took seven decades until average Englishmen saw the benefits of mechanized industry trickle down. And as many people experienced a reversal of fortunes, they vehemently opposed the introduction of new machinery. The "Luddites", as we all know, raged against the machine. It was only as workers gradually acquired relevant skills that they began to see the benefits of technological progress, and opposition ended.

A concern is therefore that this time is not all that different.

References

Frey, C. B. (2019). *The Technology Trap: Capital, Labor, and Power in the Age of Automation*. Princeton, NJ: Princeton University Press.

Frey, C. B., Berger, T., & Chen, C. (2018). Political Machinery: Did Robots Swing the 2016 US Presidential Election? *Oxford Review of Economic Policy, 34*(3), 418–442.

Frey, C. B., & Osborne, M. A. (2017). The Future of Employment: How Susceptible are Jobs to Computerisation? *Technological Forecasting and Social Change, 114*, 254–280.

Gramlich, J. (2017). Most Americans Would Favor Policies to Limit Job and Wage Losses Caused by Automation. *Pew Research Center*. Retrieved from http://www.pewresearch.org/fact-tank/2017/10/09/most-americans-would-favor-policies-to-limit-job-and-wage-lossescaused-by-automation/

Part IV

Possibilities and Limitations for AI: What Can't Machines Do?

11

What Computers Will *Never* Be Able To Do

Thomas Tozer

In 1948, John von Neumann, a father of the computer revolution, claimed that for anything he was told a computer could not do, after this 'thing' had been explained to him precisely he would be able to make a machine capable of doing it. Many scientists and philosophers strongly rejected this. For, they responded, there was something unique to humans that neither von Neumann, nor anyone else, would ever be able to replicate in a computer. Namely: consciousness. Were they correct?

Not according to proponents of Weak AI. These theorists claim that a suitably programmed computer could *imitate* conscious mental states, such as self-awareness, understanding or love, but could never actually experience them—it could never be conscious, and hence it could never be self-aware and would never actually *understand* or *love* anything. Proponents of Strong AI believe the opposite. They claim that a computer could, given the right programming, possess consciousness and thereby experience conscious mental states.

This title is inspired by Dreyfus's (1992) 'What Computers Still Can't Do'.

T. Tozer (✉)
Centre for Global Studies, London, UK

What exactly is consciousness? The philosophical literature generally adopts something on the lines of Thomas Nagel's definition: an organism has conscious mental states if, and only if, there is something it is [uniquely] like to be that organism. Nagel's example is what it is like to be a bat (Nagel 1974). Thus, a unique, subjective experience is taken to be the necessary and sufficient criterion for consciousness. Any subjective experience, whether of pain or pleasure, emotional or intellectual—that is, *all* our experiences—is an example of *consciousness*.

Could computers become conscious in the future? If the mind itself is nothing more than a highly advanced machine—the machine of the brain—as argued by philosophers such as Dennett (1991), then there is no reason in principle why it will not one day be possible to build a machine that mimics the brain and all its functions. The truth of such a view hinges on what Dreyfus coined the 'biological assumption' (1992: 156, 159–160) and the 'psychological assumption' (1992: 263): the brain is like a very advanced computer, and the mind like computer software; and mental activity can be reduced to symbol manipulation, consisting of no more than a device operating on bits of information according to formal rules. If these assumptions are correct then we should expect a sufficiently advanced computer to be able to manipulate symbols in the same way as the brain, and thereby to be able to produce consciousness.

Searle (1980) produced a powerful challenge to this view with his 'Chinese room argument': An English speaker is in a room, holding an English instruction manual on how to manipulate Chinese symbols so as to reply to questions posed in Chinese by people outside the room. He is surrounded by boxes of these symbols. By following the instructions, the man is able to pass out symbols which are correct answers to the questions and which are indistinguishable from the answers that would be given by an actual Chinese speaker. This is analogous to the symbol manipulation of a digital computer: the instruction manual represents the computer program, the questions the inputs, the boxes of symbols the database and the answers that are passed out the output.

The man thus passes the test of intelligence proposed by Turing (1950), who argued that a computer can be called intelligent if it can engage in conversation in a way that would pass for the natural language of a human being.[1] And yet the man does not actually *understand* anything of the questions he is asked or the answers he gives, thus proving, Searle argues, that computation by itself is not sufficient for understanding. Searle (1993) later extends this conclusion to consciousness, stating that the Chinese room argument shows the computational model to be insufficient for consciousness, and syntax to be insufficient for semantics. In other words, although a machine may be able to *imitate* consciousness through advanced symbol manipulation according to formal rules, this (which is all that computers are capable of) will never amount to consciousness.

Kurzweil (2005: 458–469) disagrees. He regards Searle's Chinese room argument as tautological: Searle concludes that a computer could never 'understand' anything only because he has already assumed that it is only biological entities which could be conscious and understand things. For even if no individual component is conscious, that is no reason to suppose that the *collective* of all these components, when built to the brain's level of complexity, could not be conscious. And that, Kurzweil contends, is precisely what happens in the case of the human brain: no single neuron is conscious, but when put together consciousness arises as an emergent property from complex patterns of neuronal activity. Why, he asks, could the same not happen with a sufficiently vast equivalent of the Chinese room? That is, if there were billions of people inside a massive room simulating the different processes of the brain, why should we not say that such a system is conscious?

Yet this is implausible. Of course, the collection of people may seem to act as if it has a 'mind of its own'. That is a well-known sociological phenomenon (see e.g. Le Bon 1896). But such a 'mind' is surely no more than metaphor—would Kurzweil seriously claim that a singular subjective mental experience would arise from the collective of all these people? And if so, at what point would consciousness arrive? When there were ten people? A thousand? A million? Since the *existence* of consciousness is not a graded thing, it would have to suddenly appear when there were enough people together; or it would have to be already present, if in a more subtle

---

[1] It should be noted, though, that these days the Turing test has generally been abandoned as the way to test intelligence (see New Scientist 2017: 4–5, 19, 65–67).

form, when even just two were together. Both possibilities are absurd. The same absurdity would apply to the collection of parts making up the brain and to those making up the computer: neither could produce consciousness (i.e. a singular subjective experience) by virtue of being a collective.

Dennett (see e.g. 1991, 2017) takes an altogether different perspective. He claims that subjective conscious experience does not actually exist. For him, explaining consciousness means explaining it away. Dennett regards subjective experience as unscientific, because science requires *ob*jectivity—things that can be empirically observed and measured. He therefore regards subjective experience as nonsense, and concludes that it does not actually exist.

Dennett has sustained significant criticism on this point, however, from philosophers such as Nagel (2017) and Searle (1995), who object that Dennett essentially re-defines consciousness and largely ignores its subjective aspect. As Searle (1995) puts it: 'If it consciously seems to me that I am conscious, then I am conscious. It is not a matter of "intuitions," of something I feel inclined to say.' Searle (2008: 167) also objects to Dennett's claim that consciousness is outside the scope of scientific enquiry, claiming that although consciousness may be *ontologically subjective* (i.e. its nature is such that its existence can be experienced only by the individual whose consciousness it is) it is *epistemologically objective* (i.e. universally true facts can be established about it through investigation). And it is only epistemological objectivity that science requires; there is no reason why epistemically objective study cannot take an ontologically subjective entity as its object of investigation.

Moreover, Dennett's argument amounts to a privileging of (what he regards as) scientific consistency over the most fundamental knowledge of any living person: that we experience subjective mental and physical states. If we are interested in the truth, this bias is indefensible—it betrays an irrational proclivity for science. Dennett's denial of consciousness (as a subjective entity) is therefore unfounded.

Searle's view nonetheless admits of the possibility that AI will one day be conscious: he believes that our brains produce consciousness through biological processes, and so there is no reason in principle why a conscious machine could not be built. We have not managed to build such a machine so far, Searle suggests, only because we do not understand how the brain works (see Turello 2015).

Whether building a conscious machine is actually possible turns on what Chalmers (1995) has called the 'hard problem of consciousness': how could something material, such as the brain or a computer, produce subjective experience—that is, consciousness? Dennett's response would be that consciousness, as an immaterial thing, simply does not exist; but we rejected Dennett's view above. The assumption of those such as Kurzweil and Searle, who believe (albeit for slightly different reasons) that a machine could one day possess consciousness, is that consciousness could be (and, in the case of the brain, is) the product of physical processes. I will now attempt to disprove this.

My argument for why computers will *never* be conscious consists of three premises (P1, P2 and P3) and a conclusion (C) that falls out from these. The argument is as follows: (P1) All forms of consciousness, such as understanding or love, and therefore consciousness itself, are/is non-physical. (P2) A physical thing cannot produce a non-physical thing. (P3) A computer is a physical thing. (C) Therefore a computer cannot produce consciousness.

Clearly, it is the first two premises which will be in contention here. Let us start with the first. If an object is physical then it must, *by definition*, have physical properties; therefore, if an object does *not* have physical properties, it cannot be physical. But all conscious mental states lack any physical properties. Consider love. Love can be described in non-physical terms, such as a *wish* that others be happy or a *feeling* of affection towards others, but it cannot be described in physical terms because it does not have any physical properties, such as size, weight, shape, colour or texture. It would be impossible to describe love in these terms.

Furthermore, why is it that consciousness is entirely subjective rather than objectively measurable, as ordinary material objects are? The simplest and most natural explanation is that consciousness is distinct from material things: consciousness and material objects are related, but nonetheless wholly different entities. In other words, the ontological difference between consciousness and ordinary material things (as Searle would put it) is best explained by a difference in their fundamental natures: a material (physical) nature and a non-material (non-physical) nature. For unless their respective natures were fundamentally different, why would one be observable and the other not?

That there is this difference in their fundamental natures, which corresponds to consciousness being non-physical and ordinary objects being physical, also follows from a consideration of the abilities and limits of our physical sense organs—our eyes, ears, nose and so on. These organs can observe (e.g. see/hear) *any* physical thing—that is, any ordinary material object—but they can observe physical things *only*. For example given the right conditions, such as the appropriate level of light, distance, level of magnification and so on, the eyes can see *any* physical thing that is presented to them. But they cannot see *any* object that is *not* physical, no matter what conditions are assembled. Thus, the reason why we *cannot* observe *mental* contents (i.e. others', or even one's own, subjective experiences—consciousness) with any of our sense organs, and hence cannot investigate or measure it empirically, is that these contents are non-physical. Ergo, consciousness is a non-physical entity.

But what of premise two: why should a physical thing not be able to produce a non-physical thing? The reason is that something cannot produce its opposite by itself. Water, on its own, cannot produce fire; and fire, on its own, cannot produce water. Certainly, the combination of a hot frying pan, butter and a small quantity of water might lead to fire—but water, *by itself,* cannot produce fire. Similarly, a cold temperature cannot produce warmth. It could set off a chemical reaction that might lead to the production of warmth, or the particles that are cold could themselves be warmed up, but neither would be cases of cold itself producing warmth. Indeed, there is no example to be found of an entity producing its opposite by itself.[2] Therefore, it is unreasonable to suppose that the brain would be able to do so: an entirely physical thing (brain/computer) could not produce something that is entirely non-physical (consciousness).

Note that although the above examples admit of the possibility of something leading to its opposite if combined with something else, the argument of Strong AI proponents such as Kurzweil entails no such non-physical enabling substance. Such theorists suggest that a physical thing, *and that physical thing alone,* can produce consciousness. That is precisely the claim I am rejecting.

---

[2] Or if there is, let the reader pen it in writing as an objection to this premise.

Finally, let us consider an objection from Turing (1950: 445–447). Turing thinks that the entire exercise of requiring machines to be able to experience pleasure, creativity, pain and so forth, in order to count as genuinely intelligent (or, for our purposes, conscious), is spurious. For if that test were applied to humans, they would fail it: we cannot actually be sure that another person possesses conscious mental states—only that person herself knows for certain that she is conscious. But if we are not to fall into solipsism, then to avoid continual wrangling over this point (which can never be resolved absolutely) 'it is usual to have the polite convention that everyone thinks'. Thus, rather than being forced into the solipsist position, Turing suggests that we must abandon this 'argument from consciousness' in the case of machines in the same way as we do for humans.

But Turing's appeal to 'polite convention' is unconvincing. There is a straightforward reason that applies to humans, but which would not apply to computers, why we tend to reject solipsism and believe that humans actually think: since we ourself are conscious, and since others appear to be the same kind of entity as ourself, with the same sort of biology and the same sorts of experience, it is logical to suppose that they too are conscious. This would not be true of a computer, which is completely different from a human; even a robot which looked like a human on the outside would be completely different on the inside. Therefore, it is entirely reasonable to hold computers up against a rich scepticism regarding their ability for conscious thought that we do *not* hold humans up against.

Implications for Jobs

If the argument advanced thus far is sound, then computers will never be conscious: only Weak AI is possible. Let us close by briefly considering what implications this might have for jobs.

It is conceivable that computers will come to compete with or replace humans in many jobs, including those that require cognitive ability, even though computers will never be conscious. Consider language translation: while a human translator must apply effort to understand what is meant by two different languages, so that she can understand what is said in one and then work out what phrase best expresses this meaning in the other, Google Translate does not know or understand anything about language whatsoever. All it registers is the frequency of millions of possible word sequences. The program translates between French and English based on a huge quantity of examples of French-English and English-French translations that were fed into it. These examples, combined with various mechanisms of symbol manipulation, enables language translation to take place, even though the program has no understanding of the meaning being expressed in either language. So despite the fact that it cannot understand anything, it may not be long before Google Translate can rival human translators.

This sort of example leads some to suggest that there are no jobs off limits to robots. Ford (2015: 240), for example, writes that 'Even occupations that we might expect to be reserved exclusively for people would be at risk.'

That is a stretch. Although some tasks (such as language translation) can be completed without conscious mental states, others cannot. Let us consider three mental states: empathy, responsibility and creativity. These three are all impossible without consciousness, and many jobs depend upon one or more of them.

To start with empathy: because AI will never be conscious, it will never be empathetic. And no matter how good computers get at *imitating* empathy, people will know that the computer's 'care' is only mechanical with no actual empathy behind it. A machine may learn to recognize particular emotions through external indicators such as changes in one's face or tone of voice, but its empathetic *responses* would amount to nothing more than 'hot air'.

Of course, humans are very good at anthropomorphizing—many people love their teddy bears—and there are likely to be some people who are happy with robot carers. However, the point is that there will also always be *some* (and I would suspect it would be most) people who would only accept a carer who actually cares. Such jobs would be impossible for Weak AI.

Consider next responsibility. There are many roles that people would only want occupied by a responsible person, such as the roles of judges, policemen and politicians. AI may be able to help such people make decisions, but the actual decision would always have to be made by a fully responsible person who can be held accountable for it: no one would accept a completely irresponsible and unaccountable judge who blames all his bad decisions on an algorithm. The need for responsibility is also present in transportation jobs. For example it is already possible to fully automate the flight of an aeroplane, but the profoundly high stakes of flying a plane and our reluctance to fully trust an unaccountable and non-responsible algorithm means that planes still require at least two (human) pilots.

What of creativity? Consider David Cope's 'machine composer', which creates music in the style of a particular composer. This 'composer' is so effective that audiences have been moved to tears by its 'work', describing it as just like any other classical music. Does that show that jobs involving creativity are also at risk of automation?

Not quite. For such 'creativity' is entirely artificial: it is limited by the creativity of the humans whose work is programmed into the machine. AI is only as good as the data it is given—data that represents the work of humans. Hence, an AI 'composer' can only produce good music if it has good music (written, and understood to be 'good', by humans) fed into it, that it can then *imitate*: without conscious mental states, the machine could never be 'creative' and 'conceive' of or 'imagine' anything outside of what has already been produced. Therefore, machines will never be able to compete with, for example, those humans who actually create music, or who create any art that is completely new and original.

The conclusion, then, is that progress in (Weak) AI will threaten many jobs—but not all. Because computers will never be conscious, jobs requiring conscious mental states, such as empathy, responsibility and creativity, will remain impossible for computers even as AI's ability to imitate these mental states continues to race ahead.

References

Chalmers, D. J. (1995). Facing Up to the Problem of Consciousness. *Journal of Consciousness Studies, 2*(3), 200–219.
Dennett, D. C. (1991). *Consciousness Explained*. Boston: Little, Brown.

Dennett, D. C. (2017). *From Bacteria to Bach and Back: The Evolution of Minds.* London: W. W. Norton.
Dreyfus, H. L. (1992). *What Computers Still Can't Do: A Critique of Artificial Reason* (Rev. ed.). Cambridge, MA: MIT Press.
Ford, M. (2015). *The Rise of the Robots: Technology and the Threat of Mass Unemployment.* New York: Basic Books.
Kurzweil, R. (2005). *The Singularity is Near: When Humans Transcend Biology.* New York: Viking Penguin.
Le Bon, G. (1896 [2001]). *The Crowd: A Study of the Popular Mind.* Kitchener: Batoche Books.
Nagel, T. (1974). What Is It Like to Be a Bat? *The Philosophical Review, 83*(4), 435–450.
Nagel, T. (2017, March 9). Is Consciousness an Illusion? *The New York Review of Books.*
New Scientist (2017). *Machines That Think: Everything You Need to Know About the Coming Age of Artificial Intelligence* (N. Bostrom, N. Christianini, J. Graham-Cumming, P. Norvig, A. Sandberg, T. Walsh, contributors, D. Heaven & A. George, Eds.). New Scientist.
Searle, J. R. (1980). Minds, Brains, and Programs. *Behavioral and Brain Sciences, 3*(3), 417–424.
Searle, J. R. (1993). The Problem of Consciousness. *Consciousness and Cognition, 2*(4), 310–319.
Searle, J. R. (1995, December 21). 'The Mystery of Consciousness': An Exchange. *The New York Review of Books.*
Searle, J. R. (2008). *Philosophy in a New Century: Selected Essays.* Cambridge: Cambridge University Press.
Turello, D. (2015). Brain, Mind, and Consciousness: A Conversation with Philosopher John Searle. *Library of Congress*, March 3. Retrieved from https://blogs.loc.gov/kluge/2015/03/conversation-with-john-searle/
Turing, A. M. (1950). Computing Machinery and Intelligence. *Mind*, New Series, *59*(236), 433–460.

12

Possibilities and Limitations for AI: What Can't Machines Do?

Simon Colton

I will talk here about limitations and possibilities for Artificial Intelligence (AI) in the future of work. I will put forward my opinion that we can and should take a practical rather than philosophical view of what intelligent algorithms are already doing in work environments, and plan for solid and sensible, rather than sensational, mid-term progress that will bring benefits to workforces, but only if advanced AI systems are properly handled by business leaders and politicians, amongst others. I will first discuss the limitations of AI systems, and will then propose some of the possibilities for software in my particular area, Computational Creativity, where we build software to act as muses, tools, co-creators and fully autonomous creative entities (Colton and Wiggins 2012).

S. Colton (✉)
Department of Electronic Engineering and Computer Science, Queen Mary University of London, London, UK

Sensilab, Faculty of Information Technology, Monash University, Melbourne, VIC, Australia
e-mail: s.colton@qmul.ac.uk

It is worth remembering that AI research is predominantly carried out as an experimental science, very much like physics and chemistry. The majority of researchers in AI are pushing from the bottom up, that is, taking existing problems which we know AI systems can solve, and improving the underlying algorithms in a general way to produce better solutions in more efficient ways. Other researchers, like myself, pull AI up by the hair by proposing and investigating new and difficult intelligent tasks for automation, such as mathematical invention, fictional ideation or videogame creation. Most researchers separate the technology from the ethical considerations of how their technology is employed in society. Other researchers, again like me, try to be part of the cultural communities in which their software is employed, and are sensitive to the ethical issues raised by AI involvement in previously human-only activities.

It is also worth remembering that computer systems are not human. In this age of rampant anthropomorphization, many people (including some AI researchers) talk of AI systems as if they were some kind of sub-human, rather than a completely different kind of information-processing system. Often, such blind comparisons of AI systems and people lead to inappropriate projections of human attributes onto software or robots. When engaged in debate about, say, machine consciousness, I often bring up the question of automating fingernail growth, which is very useful to people, but absurd for computers. This highlights the fact that people are really very different to computers, and we don't need to simulate everything about humanity in them. In particular, software may not need anything like human consciousness to achieve great intelligence and purpose or to be of use to human society.

I reject the premise of the question: 'What can machines not do?' that has been put forward as a springboard for discussion in this forum. Of course, there are many aspects of human intelligence that have little or no computational equivalent in today's AI systems. But I see no fundamental reason why software won't have human-like (or super-human) intelligence and life-like features in the future. We will likely have difficulty embedding things like a 'soul' in computers, but I expect this will be because we can't easily define such terms for people, let alone artificial entities. To my mind, human-centric terms such as creativity and consciousness will eventually yield to scrutiny via more holistic studies, and just as we are beginning to accept that people are not the centre of the intelligence universe, we will eventually have generalized notions of creative or conscious behaviour, and computational equivalents thereof.

Engineers such as myself listen to well-meaning arguments from non-experts about, say, how software will never be creative, consider the issues raised, for example about lack of intentionality in computers, and then write code and produce autonomous systems that basically invalidate those arguments. Such debate is a great driving force for AI research, but I believe it is naive to think that there are fundamental technical limitations on the levels of intelligence that software can achieve, if society wants it to. However, while I am very much a techno-supporter, I am also a realist in terms of the timescales which will be needed to achieve the kind of systems described above, and in terms of society's willingness to accept AI systems exhibiting qualities which are so dearly held as defining characteristics of humanity.

My brother and I were born shortly after the 1969 moon landings. We grew up believing in absolute terms that we would be travelling to the moon and beyond for our holidays. Given the current crazy levels of hype over the power of AI systems, my three-year-old daughter may well grow up believing that her best friend in university will be an android. Sadly, she, like her uncle and me, will be disappointed. While there has been a step change in the power of AI systems, brought about in the last decade by advances in deep learning techniques, AI systems are not nearly as intelligent as the press, politicians and philosophers would like us to believe. The hype is understandable: technology leaders have to hugely overstate the life-changing power of their AI systems to have any chance of gaining venture capital these days; journalists have to overstate the strength of results from AI projects, to compete in a clickbait environment; and in order to make a name for themselves, politicians and philosophers need to take an extreme and short-term view of AI in order for it to appear relevant and timely.

With larger proportions of the world's smartest people working on AI development recently, the pace of advance has certainly increased, and I sincerely hope that the benefits from truly intelligent AI systems come sooner rather than later. However, the evidence from the history of AI is that progress is slow and hard fought. It is sobering to think that the last time journalists were talking about AI systems playing board games like chess and Go, was in the Kasparov/Deep Blue era more than 20 years ago. And, with all due respect, board games are not particularly important—they are pastimes for people and don't deserve to be spoken about with the same kind of reverence as, say, universal physical laws, as some AI researchers do. We (human society) love chess and Go, precisely because they are very difficult for people: it takes huge dedication to master a board game and this really separates the masters from the novices, which we prize in competitive human society. We (AI researchers) love chess and Go, precisely because playing them is a relatively easy activity for software: given the closed world, simple rules and transparent nature of the competition, they are ideally suited to AI-style search techniques, and indeed games continue to be a huge driving force for our field.

Hence we should take a more realistic look at recent breakthroughs in AI, for instance the super-human Go playing abilities exhibited by the AlphaGo Zero system from Google DeepMind (Silver et al. 2018). While it's a huge achievement, especially as the software learns to be a grandmaster from scratch by repeatedly playing against itself, we should not extrapolate too far from this milestone being reached. Importantly, of course, this level of super-human intelligence is not likely to negatively impact the world of work. Human chess champions, for instance highlight the humanity in the heroic battles of two people over a chessboard, and are not perturbed much by the fact that there is software that can beat them. Moreover, AI spectacles such as AlphaGo beating the world Go champion serve to make the game more popular, rather than less. For instance, the world apparently ran out of Go boards to sell, shortly after the AlphaGo event (Shead 2016).

A more sensible view of AI abilities, which is held by the majority of practitioners in the field, should extend to projections of software counterparts taking over from human lawyers, doctors, scientists, journalists, and so on, speculation of which is rife across the media and in some academic circles. Such speculation has been fuelled by books like Nick Bostrom's on SuperIntelligence (Bostrom 2014), where he is clear that his philosophical enquiry is entirely speculative, except in one respect: that AI superintelligence can and probably will occur in lightning fast time, for example overnight. While this is a brilliant and much-used science fiction meme, it is, unfortunately, bad fictional science. No AI researcher I know has the first clue about how we could achieve overnight superintelligence, and as far as I know, no-one has a reasonable road-map for so-called Artificial General Intelligence, with metrics for partial progress remaining controversial and problematic (Goertzel 2014).

It is worth debunking a couple of Bostrom's ideas on how such rapid superintelligence could be achieved. One is incremental automated AI engineering, that is an AI system writing a slightly more intelligent AI system, which itself writes an even more intelligent system, and so on. This is basically the wet dream of undergraduates embarking on an AI course. Sadly, there is a chicken-and-egg problem not addressed here. In reality, you need human-level intelligence to engineer very stupid AI software. Very stupid AI software is not going to produce slightly more intelligent software that a very clever AI researcher could not. I'm actually working myself on AI systems which perform automated software engineering, and I'm confident that AI systems can write code of real practical value. I'm also confident that someone will solve the chicken-and-egg problem, but I don't have a solution, and I suspect one will be a long time coming and rely on many other breakthroughs in the science and engineering of AI systems.

Another approach put forward by Bostrom is the automated evolution of brain-like structures equivalent to those in people's heads. He discusses calculations involving vast numbers to help guess roughly how long it would take to evolve something equivalent to a brain. But this misses the important fact that brains (and the rest of our human bodies) evolved within natural environments where, in general, only the fittest individuals survived to pass on their genetic information. It's not even clear what kind of environment (real or simulated) would be required to evolve superintelligence, but it is clear that the test of fitness would be long and very difficult, requiring geological rather than technological timescales. Advances in evolutionary AI systems happen each year, but it is difficult to imagine the automated evolution of a human-like brain structure in any reasonable timescale that should influence our discussion of AI in the workplace, unless we want to engage in science fiction, as Bostrom does.

To summarize so far, I believe that any technical limitations on the abilities of AI systems to become intelligent enough for employment in workplaces instead of people are not in terms of fundamental, theoretical issues, but rather in terms of the speed to make scientific discoveries about computational intelligence, and in terms of engineering systems to take advantage of these breakthroughs. While it would not surprise me if we saw autonomous cars routinely on our streets in 10 years' time, it would also not surprise me if it took another 50 years for this to happen (Tschangho 2018).

The other limitations on the usage of AI in the workplace, will I hope, be self-imposed, as society in general responds to automation. Slowly but surely, AI systems will gain abilities to take on the duties of people undertaking intelligent tasks. To highlight what I see as the main issue with this, I use a thought experiment called: 'a new kind of lucky', as follows. Albeit begrudgingly, we accept different kinds of lucky people in society, for instance lottery winners, or people born to wealthy parents and so on. In an age of automation, imagine a company boss announcing that 'this week's lucky person is Joan Smith in accounts, because her job has been automated'. The lucky winners in this scenario are then given the option of going on permanent leave with a full salary, continuing to do aspects of their current job, or transferring to a new, suitable and presumably more interesting role in the company. While there may be an overhead in buying software to automate the work of an employee, the overhead of running it (for 24 hours a day, of course) would be low, and the financial impact of paying the lucky employee would likely be relatively neutral.

The thought experiment ends by considering what is wrong with this scenario, and it's easy to see that the so-called lucky person would actually be very unlucky because, in current our current capitalist society, they would be sacked, with any money saved used possibly for reinvestment, but equally likely so that the company bosses and shareholders could hoard more money. We could, of course, campaign as a society for ethical roll-out of automation, and press our politicians to enshrine this in legislation. However, quite the opposite seems to be happening. Some taxi companies for example seem to be using human 'labour purely as a stop-gap to raise venture capital for research and development, so that in the longer-term they can roll-out a fleet of autonomous cars, eventually putting their entire human workforce out of a job (Price 2019).

I hold quite a utopian view that automation can free humanity from the drudgery of meaningless toil—which, taking a sincere look at the world of work—is what many people in paid employment are asked, or indeed forced via circumstance, to do. This may be a naive position to hold, and certainly has numerous issues. In particular, it is clear that long-term benefits to society from automation will bring short/medium-term difficulties in job losses, with accompanying loss of earnings, self-esteem and well-being. As a society, we may want the long-term benefits, but not be prepared to insure against the short-term difficulties. Can we afford to implement a three-day working week, instigate a universal income or increase national insurance to enable unemployed people to live good lives, when we can barely care properly for the sick or educate the young (as the rhetoric goes)? Maybe none of these measures is necessary if we can soften capitalist values and carefully control the usage of AI systems in the workplace.

While we usually turn to scientists and technology leaders, we may also look to the arts for guidance with respect to our technological future. I have been involved in cultural projects where AI-generated fictional ideas have become the basis for a West-End musical theatre production (Colton et al. 2016); AI-generated paintings have been exhibited in art galleries and museums (Colton and Ventura 2014); AI-generated videogames have been entered into competitions and eagerly played (Cook et al. 2016); and AI-generated poems have been read out on BBC radio (Colton et al. 2012). This has enabled me to look at how the art world is changing to accommodate artists who use computers in their work, artists who hand over some level of creative responsibility to software, and indeed autonomous AI artists, such as my software called The Painting Fool (Colton 2011). This has led to an understanding of the merits and drawbacks of using AI systems in human-centric areas, with notions possibly transferable from the arts to the wider world of work.

Within both academic and cultural contexts like those mentioned above, I use another thought experiment to highlight upcoming issues with my field of Computational Creativity, where we study how to hand over creative responsibility to software in arts and science projects. In particular, I read out the following short poem entitled 'Childbirth' and discuss with the audience what the female author may have been trying to express.

> The joy, the pain, the begin again. My boy.
> Born of me, for me, through my tears, through my fears.

I then inform the audience that the author was actually a man, and this changes the perception of the author's intention for some people. I then drop the bombshell that the man was a convicted paedophile, and the interpretation of the childbirth poem gets somewhat dark. When I change the authorship again to be a computer program run with little human guidance, I ask the audience to think about what the software 'meant' with the poem, and we usually agree that something is now missing: the software knows nothing about childbirth and so the poem is somewhat inauthentic, the words are maybe meaningless and the poem perhaps pointless. Finally I point out the truth that I wrote the poem, and found it remarkably easy to pen some words which change radically in meaning as knowledge of the authorship and hence authenticity shifts.

If it isn't already, I believe authenticity will take centre stage in the arts as the artistic output from AI systems gains in quality, and the backstory surrounding the creative process increases in intrigue (Colton et al. 2018). In one mode, for example, The Painting Fool paints portraits in a simulated emotional response to newspaper articles it reads in the Guardian (Colton et al. 2015). The backstory from the Guardian and the software's response, as well as the experience of posing and seeing the portrait emerge through a simulated hand and simulated paint strokes on screen, are usually more impressive and interesting to the sitters than the artwork produced. That said, there are limitations on the backstory and the simulation of emotions that The Painting Fool can exhibit. While I don't believe we are heading towards some kind of 'singularity' where people and computers merge, it will be possible to give software increasingly human-like qualities, and we may come to appreciate the authentic life experiences that computers/robots have, as expressed through their art. This doesn't mean, though, that robots will ever get over the fundamental limitation of being non-human, and I believe this will set their art apart from that of people, and society will realize that at its best, art is actually a celebration of humanity.

There is a current fascination with AI-generated artworks, as there usually is when new technologies are used in the arts: the so-called shock of the new has much cultural merit when presented appropriately. In response, people are already drawing up boundaries for what automation could and should be used for. Naturally, those boundaries will be crossed and blurred, and the world will be a more interesting place as a result. However, art is a particularly human-centric field, and this will surely become clearer in an age of automation. I have occasionally described poems, for instance as condensed humanity: written by people, for people and usually about people (Colton et al. 2012).

Returning to consideration of work activities, there are certain things analogous to poetry that we might want to ring fence as desirable for human-only or human-centric undertaking. One of these is elderly care work, which has been in the news recently, as robotic carers are going from science fiction to science reality, with test cases now in Japan and elsewhere (Hurst 2018). In such domains of work, we will have to determine a balance to try to achieve. With automation of other roles, there may be more people available for care work, which will be necessary due to improvements in healthcare leading to ageing populations. Yet, while it's fairly easy to argue why human rather than robotic interactions are desirable in elderly care, there is surely a utilitarian argument that robotic care is better than none, and a balance of both is a good idea. In other areas, there will be other considerations which override any kind of human-only ring fencing. I used to remind students that spreadsheets were, long ago, paper ledgers completed by hand by people good at arithmetic and statistics. However, computerized versions were implemented and made widely available partly to democratize bookkeeping, partly to increase the accuracy and power thereof, and partly to extend the usage of spreadsheets into many other aspects of life. While human accountants are still going, it would be strange to think of someone employed to do spreadsheets, or any complex calculations by hand these days.

I heard on the radio a simple but powerful definition of work and leisure that has influenced my thinking about how AI could be deployed in society. Work was defined as any activity that you would gladly pay someone else to do, with leisure defined as everything else. This definition was used in an operational way, with researchers asking people which of their daily activities they would classify as work, and which as leisure. Moreover, the notion of paying someone to undertake a task sets quite a high bar for what is considered work, and gets people thinking about the quality of their work and leisure lives. It is, therefore, quite a subjective definition. For example recreational fishing is absolutely a leisure activity for many people, but I would gladly pay someone to sit and hold a rod for hours on end, which would feel like work to me. I have done my fair share of nappy changing, but it was for my own baby and I wouldn't choose to pay someone to do it, not because I like doing it, but because it personally doesn't seem worth paying someone for this, although I respect other people who do pay. I am lucky in that I would class most of my paid employment as leisure and even more lucky that my employers don't exploit this fact.

It would seem to me that a good place to start in the roll-out of automated systems in the workplace is to ask the employees which of their activities they would categorize as work, and which as leisure, then attempt to automate the work parts, and increase the opportunities for leisure activities that they undertake. An example I've used in the past is clickbait journalism. An ex-employee was interviewed for the Guardian (Anonymous 2016), and said that he disliked his job as a clickbait journalist, having to knock out dozens of salacious and vacuous short articles a day. While this particular journalist chose to move on, he could have been replaced by automation, as software systems are beginning to achieve high standards in rehashing statistics and third party news stories into clickbait articles. A third way would have been possible: the journalist could have told their employer that they have really enjoyed penning the occasional human interest piece and agree with them to publish more in the time freed up by the software taking on responsibility for articles about celebrity pet makeovers and the like. As mentioned above, though, this is currently an unlikely ending to the story.

While I know it is desperately utopian, I still believe that we can live in an automated world where both human-driven and autonomous cars drive us around cities, but the human drivers are those who want to be there because they truly love their job (and not just the economic security it brings). We may pay more for the human experience, as we currently do when buying works of art. The extra payment will be necessary as people cost more to employ than robots, but we will gain more culturally from interaction with the driver, and probably enjoy the experience more. This utopian view naturally doesn't extend to all types of jobs, people or companies, but it is, I believe, a maxim to work towards as a society.

In the forum which led to this essay, there were a number of issues raised by the other participants, some of which I have tried to address above. In addition, concerns were raised about my point that the majority of AI researchers tend to separate development of AI technology from its uses and not worry too much about ethical issues. I emphasized that while this is currently the case, things are changing, with technology leaders such as Demis Hassabis from Google Deep Mind promoting ethical usage of AI, and ethics courses being given to computing students.

Later, we returned to the question of who makes the decisions about AI usage, and discussed whether this is likely to come from the bottom up, for example from community or consumer groups, and I confessed to being dubious about this. I pointed out that the academic AI community is also leading the way, giving as an example a movement against autonomous killing machines spearheaded by an old colleague of mine, Prof Toby Walsh (Walsh 2018). When I expressed scepticism to Toby recently about whether an academic movement could alter government, military or even commercial direction, he was more optimistic than me, and he also pointed out that such organizations tend to hire PhD graduates, and competition is very tough for the brightest sparks. Some of these graduates, he continued, may be influenced by academic movements, and thus take into account the ethical position(s) of the organizations looking to hire them, which is a valid and optimistic point.

Another line of discussion focused on the notion of software itself being creative. It was suggested that while software can produce novel combinations of previous great works, say, in musical composition, artists don't just draw on datasets of music, but also impressions from their daily lives, from other cultural media such as novels, from dreams, from interactions with friends and many other sources. Only with these other influences would software be able to create great art, it was argued. This is a position I also hold, and I often talk about how we should avoid pastiche generation in Computational Creativity research, as it's not something often associated with highly creative behaviour. However, I reject the implied position that software creativity can't be influenced by all the other sources mentioned. I pointed out that The Painting Fool is inspired by daily newspaper articles and twitter streams, and was previously inspired by the emotions being expressed by sitters in portraits. I re-iterated that I don't see any reason why software can't have rich sources of influences, as analogies to human life can be engineered into software. I speculated that the reason this isn't currently the case in creative software is not because of technical limitations, but rather that there is no obvious economic model yet for computer creativity, which might mobilize the vast human and computational resources of the big technology companies.

Finally, the question of evaluation came up, and we discussed creative output being original and having value. It was pointed out that many creative people, all of whom claim to be original, do badly on the market, and this could be a matter of talent, scarce resources, biased evaluation and so on. Evaluation is a very noisy process, and we questioned whether a computer could engage in such a noisy environment, perhaps learning about themselves, and improving from their failures and errors, as people do. I noted that The Painting Fool does indeed learn from its (relative) failures to become slightly better at achieving affective portraits each time it perceives a failure. It does this using machine vision to analyse its output and project emotional value, with the simulated emotion coming from reading newspaper articles.

I also debated whether we should be reliant on simplistic definitions of creativity in the literature such as producing novel and valuable outputs. I put forward the point of view that notions such as 'art' and 'creativity' are actually essentially contested concepts (Colton et al. 2014). These are defined as concepts for which the proper usage involves endless debate about their proper usage (Gallie 1955). In other words, we have chosen as a society to agree to disagree forever about certain notions, as this is an important driving force for progress. The following discussion centred on whether certain people had more authority in discussions about essentially contested concepts such as art and creativity. I agreed that some people do have more authority and understanding, and therefore may be more convincing in their arguments, but that doesn't change the fact that to properly discuss creativity, they should be arguing with someone about it. This led to the close of the discussion, where I pointed out that I would only believe I had succeeded in producing truly creative software if it could itself contribute to the debate about what creativity is. I suggested that engineering such a philosophically creative AI system would provide a suitable end to my career as a Computational Creativity researcher.

References

Anonymous. (2016, November 21). The Secret Life of a Clickbait Creator: Lousy Content, Dodgy Ads, Demoralised Staff. *The Guardian*.
Bostrom, N. (2014). *Superintelligence: Paths, Dangers, Strategies*. Oxford University Press.
Colton, S. (2011). The Painting Fool: Stories from Building an Automated Painter. In J. McCormack & M. d'Inverno (Eds.), *Computers and Creativity*. Springer.
Colton, S., Cook, M., Hepworth, R., & Pease, A. (2014). *On Acid Drops and Teardrops: Observer Issues in Computational Creativity*. In Proceedings of the 7th AISB Symposium on Computing and Philosophy.
Colton, S., Goodwin, J., & Veale, T. (2012). Full-FACE Poetry Generation. In *Proceedings of the Third International Conference on Computational Creativity*.
Colton, S., Halskov, J., Ventura, D., Gouldstone, I., Cook, M., & Pérez-Ferrer, B. (2015). *The Painting Fool Sees! New Projects with the Automated Painter*. In Proceedings of the Sixth International Conference on Computational Creativity.
Colton, S., Llano, M. T., Hepworth, R., Charnley, J., Gale, C. V., Baron, A., Pachet, F., Roy, P., Gervás, P., Collins, N., Sturm, B., Weyde, T., Wolff, D., & Lloyd, J. R. (2016). *The Beyond the Fence Musical and Computer Says Show Documentary*. In Proceedings of the 7th International Conference on Computational Creativity.
Colton, S., Pease, A., & Saunders, R. (2018). *Issues of Authenticity in Autonomously Creative Systems*. In Proceedings of the Ninth International Conference on Computational Creativity.
Colton, S., & Ventura, D. (2014). *You Can't Know My Mind: A Festival of Computational Creativity*. In Proceedings of the 5th International Conference on Computational Creativity.
Colton, S., & Wiggins, G. A. (2012). Computational Creativity: The Final Frontier? *Proceedings of the European Conference on Artificial Intelligence*.

Cook, M., Colton, S., & Gow, J. (2016). The ANGELINA Videogame Design System, Parts I and II. *IEEE Transactions on Computational Intelligence and AI in Games, 9*(2/3), 1–13.

Gallie, W. B. (1955). Essentially Contested Concepts. *Proceedings of the Aristotelian Society, 56*, 167–198.

Goertzel, B. (2014). Artificial General Intelligence: Concept, State of the Art, and Future Prospects. *Journal of Artificial General Intelligence, 5*(1), 1–26.

Hurst, D. (2018, February 6). Japan Lays Groundwork for Boom in Robot Carers. *The Guardian*.

Price, R. (2019, April 12). Uber Says Its Future is Riding on the Success of Self-driving Cars, but Warns Investors That There's a Lot That Can Go Wrong. *Business Insider*.

Shead, S. (2016, April 4). There's a Worldwide Shortage of the Board Game Go after Google's Computer Beat the World Champ. *Business Insider*.

Silver, D., Schrittwieser, J., Simonyan, K., Antonoglou, I., Huang, A., Guez, A., Hubert, T., Baker, L., Lai, M., Bolton, A., Chen, Y., Lillicrap, T., Hui, F., Sifre, L., van den Driessche, G., Graepel, T., & Hassabis, D. (2018). Mastering the Game of Go Without Human Knowledge. *Nature, 550*, 354–359.

Tschangho, J. K. (2018). Automated Autonomous Vehicles: Prospects and Impacts on Society. *Journal of Transportation Technologies, 8*, 137–150.

Walsh, T. (2018). *Machines That Think: The Future of Artificial Intelligence*. Prometheus Books.

Part V

Work in the Digital Economy

13

Work in the Digital Economy

Daniel Susskind

In this talk, I want to do two things. First, I want to offer a very brief intellectual history of the way in which many economists have thought about technological change and its impact on the labour market in recent decades. Then I want to use these ideas—and their limitations—to set out a few implications for the future of work. This talk draws explicitly on other work I have done: for instance two books, *The Future of the Professions* (2015/2017) and *A World Without Work* (2020), other pieces of research, and a recent TED Talk, '3 Myths about the Future of Work (and Why They're Not True)'.

I want to begin with the strange changes that took place in labour markets from the 1980s to the turn of the twenty-first century. During that period, if you had lined up workers in many countries from lowest-skilled to highest-skilled, you would have found that low-skilled and high-skilled employment shares at either end of the line grew, but employment shares for those in the middle shrunk (by 'employment shares', I mean the share of these roles in overall employment). Economists call

---

D. Susskind (✉)
Balliol College, University of Oxford, Oxford, UK

© The Author(s) 2020
R. Skidelsky, N. Craig (eds.), *Work in the Future*,
https://doi.org/10.1007/978-3-030-21134-9_13

this the 'polarisation' or the 'hollowing-out' of the labour market. Typically, people point to it unfolding in the US, but the general picture appears in many other countries too.

This presented economists with an empirical puzzle that was quite different from the one that had preoccupied them in the past. Previously, the focus for many had been on why the so-called 'skill-premium' was rising; why, for large parts of the twentieth century, the wages of high-skilled workers were rising relative to the wages of low-skilled workers, even though the supply of the former was rising as well. (Typically, the premium was measured by comparing the wage of college graduates to those with only a high school education.) There was a popular explanation of this: technological change was 'skill-biased' and, for various reasons, new technologies raised the demand for high-skilled workers relative to low-skilled workers, pushing up their wages and that skill-premium. But the trouble was that this story could not explain the hollowing-out of the labour market. There, the interesting fact was not so much that high-skilled workers were increasingly being paid more relative to low-skilled workers, but that those middling-skilled workers were enjoying neither the same wage nor job growth as the low-skilled or high-skilled at either end.

In light of this shortcoming, support began to build for a different narrative, developed by a team of economists at MIT: David Autor, Frank Levy and Richard Murnane. This was known as the 'Autor Levy Murnane hypothesis'; the 'ALM hypothesis' for short. This story has two distinct parts to it. The first is that thinking about the labour market in terms of jobs is misleading; instead, we need to think in terms of all the different tasks that make up any particular job. When we think about the future of work, we tend to slip into a 'jobs' mindset, talking in terms of entire roles. What does the future look like for 'doctors' and 'lawyers', 'teachers' and 'accountants', 'architects' and so on? But this is unhelpful, because the really interesting churn takes place at the level of the different composite tasks that make up any job. Focusing on high-level 'jobs' masks these interesting, subtler, changes.

The second part of the thesis is a distinction between 'routine' tasks and 'non-routine' tasks. 'Routine' tasks are those the performance of which human beings find easy to explain; 'non-routine' tasks are those that that human beings find difficult to explain. The distinction rested on the work of Michael Polanyi, the Hungarian philosopher. He drew a distinction between 'tacit' and 'explicit' knowledge. And the ALM hypothesis argued that 'routine' tasks require 'explicit' knowledge, the sort of knowledge that human beings find it easy to articulate and, conversely, 'non-routine' tasks require 'tacit' knowledge, the sort of knowledge that human beings find difficult to articulate.

This two-part approach—think in terms of tasks and distinguish between 'routine' and 'non-routine' ones—fitted neatly with the conception of machine capabilities that many economists held at the time that the ALM hypothesis was being developed. They thought that if you wanted to automate a task, you had to sit down with a human being, get her to explain how it was she performed a particular task, and then try to capture that explanation in a set of rules or instructions for a machine to follow. Clearly, for 'routine' tasks this would be straightforward; human beings could explain how they perform these tasks with comparative ease, and it would be easy to write a set of instructions based on that explanation—so the task could be readily automated. But for 'non-routine' tasks, that would not be possible—and so they would be hard to automate.

It was not only economists who thought about machines in this way: the spirit of this distinction between 'routine' and 'non-routine' tasks was once popular in the field of artificial intelligence, too. I know this because my co-author in writing *The Future of the Professions*, and my father, Richard Susskind, wrote his doctorate during the 1980s on artificial intelligence and the law at Oxford. Even back then, he was trying to build systems that could solve legal problems. To see what he was doing, consider one example of his work. In the late 1980s, a complex piece of legislation was passed in the UK called the Latent Damage Act 1986. At that time, the leading expert on this area of law was a man called Phillip Capper, who happened to be the chair of the law school at Oxford. And he came to my father and said something along the lines of, 'It is absurd. When someone wants to understand the impact of this complex legislation, she has to consult an expert, and there are almost none in this field. Instead, why do we not work together and build a system based on the expertise that I have in my head so non-lawyers can tap into this without speaking to an expert?' That is what they did, developing what became the first commercially available 'expert system', as they were known then, in the law. Users answered 'yes' or 'no' to a set of questions and the system would then give users an answer to a difficult question—when does my action run out of time? This approach to building a system was precisely the picture that economists who adopted the ALM hypothesis had in mind: in order to automate a task, you had to sit down with someone (in this case, Phillip Capper) get him to explain to you how it was he performed it, and then you built a system, a representation of that explanation, for non-experts to use.

What was so compelling about the ALM hypothesis, from an economic point of view, was that it could explain the hollowing-out of the labour market that was taking place—unlike the theory of skill-biased technological change. When, following the thesis, economists took jobs from across the labour market and broke them down into all the tasks that made them up, it transpired that many high-skilled and low-skilled jobs required 'non-routine' tasks—that was why they were hard to automate, and why they saw great employment growth. But, critically, those middling-skilled jobs were disproportionately composed of 'routine' tasks—that was why, in contrast, they could be automated with relative ease, and why their employment share fell.

For a while, this was the dominant way that economists thought about how technological change affected the labour market—there was a realm of 'non-routine' tasks at either end of the labour market, out of reach of machines, and left exclusively for human beings to do. But in the last few years, a problem emerged. In my own research, I asked the question: what do the tasks of driving a car, making a medical diagnosis and identifying a bird at a fleeting glimpse have in common? And the answer is that they are all tasks that, until recently, most economists thought were 'non-routine' and so could not readily be automated. However, now they increasingly can be. Today, all major car manufacturers have driverless car programs, there are countless systems that can diagnose medical problems, and there is even an app developed by the Cornell Lab of Ornithology that can identify a bird at a fleeting glimpse.

What went wrong with the ALM hypothesis? As we have seen, most economists thought that the only way to automate a task was to copy the way that human beings thought and reasoned, to try to capture their thinking and reasoning processes in a set of instructions for a machine to follow. When my father was writing his doctorate, that view may have been right; but today, it is problematic. This is no longer how these latest systems and machines work. Advances in processing power, data storage capability and algorithm design, mean that the 'routine' verses 'non-routine' distinction is far less useful than it was before.

To see why, take a concrete example: the task of making a medical diagnosis. This is a classic case of a 'non-routine' task—a doctor would struggle to articulate exactly how she makes a diagnosis, find it hard to pinpoint the precise rules and thinking processes she goes through in reaching a decision. Yet a system was recently developed at Stanford that can tell whether or not a freckle is cancerous as accurately as leading dermatologists. How does it work? It is not trying to copy the doctor's reasoning processes. It 'knows' or 'understands' nothing at all about medicine. Instead it has a database of about 129,450 past cases and is running a sort of pattern recognition algorithm through them, hunting for similarities between those images and the photo of any troublesome lesion under scrutiny. It does not matter that this task is 'non-routine', that a human being might not be able to explain exactly how she makes a diagnosis—this system is performing the task in a very different way, an unhuman one, based on the analysis of more possible cases than a doctor could hope to review in her lifetime.

Again, this strikes closer to home. For many people, a turning point in artificial intelligence came in 1997, when Garry Kasparov, then the world chess champion, was beaten by Deep Blue. Intriguingly, if you had gone back in the 1980s and asked my father and his colleagues if they thought it would ever be possible to build a machine that could beat a chess champion like Garry Kasparov, they would have said 'no'—and remember, these were some of the most progressive people working on these technologies at the time. And the reason they would have said no would have followed the reasoning we have seen before. They thought the only way to build a high-performing system was to sit down with a human expert, get them to explain to how they solved a problem and then try to represent that in a set of rules, like a large decision tree. But here was the problem: if you sat down with Garry Kasparov and asked, 'Garry, how are you so good at chess?', he might be able to give you a few opening moves and closing plays, but ultimately, like doctors, he would struggle to articulate any underlying rules. In all probability, he would say something like 'it's gut reaction, instinct, intuition'. Diagnosis and game playing at a high level, in short, would be regarded as 'non-routine', and so, they imagined, the underlying expertise was not reducible to a set of rules that a machine could follow.

In retrospect, AI specialists of that era now realise their mistake. They had not expected the exponential growth in processing power that would come about in the years that followed. By the time Garry Kasparov played Deep Blue, that machine was calculating up to 330 million moves a second. Garry Kasparov, at best, with a following wind, could ponder maybe 110 in his head during any one turn. Garry Kasparov was blown out of the water by brute-force processing power, advanced software, operating on vast bodies of data. In a sense, Deep Blue was playing a different game. It was not trying to copy the rules Kasparov followed or the thinking process in which he was engaged. Since then, many systems and machines have been built in this way. And, as a result, many 'non-routine' tasks are within reach of machines.

So what are the implications for thinking about the future of work?

The first concerns the limits of machine capabilities. It is now clear that, as a result of advances in processing power, data storage and algorithm design, machines are able to perform tasks in fundamentally different ways to the way that human beings do. This means that our lack of understanding about human intelligence—about how we think and reason, about the nature of consciousness and the mind—is far less of a constraint on automation than it was thought to be in the past. In this talk, I have set out one example of this: a traditional bottleneck to automation—the inability of human beings to articulate how the rules and thinking processes they engage in when performing a 'non-routine' task—matters far less. More broadly, it means that, in thinking about the future of work, the important questions are not the philosophically fascinating ones, like 'can a machine ever be conscious?', but the more bluntly practical ones, like 'can a machine ever perform a task that requires consciousness in a human being, but by doing so in a different way'?

The second implication concerns the pervasiveness of automation. It is also now clear that the impact of technological change in the future is likely to be felt right across the labour market. Again, one of the unhelpful things we do when we talk about work is that we talk about the different jobs that people do: and instead, as I have explained, what we need to do is think in terms of tasks. When you do that, you find two things: first very few existing roles can be fully automated given emerging technologies, but also that almost all jobs involve tasks that can be automated. There is other compelling research that confirms this observation. The risk, then, if you continue to think in terms of 'jobs' alone, and only keep an eye out for roles that are likely to be taken on in their entirety by machines, is that you will seriously underestimate the impact of technological progress on the labour market.

The third implication concerns what I call the 'skill-blindness' of technological change. There is a presumption, reinforced by the traditional skill-biased view of technological change, that technology tends to help workers with skills and harms those without. But what has become clear is that, actually, technological change is not necessarily biased towards particular types of workers at all, but is biased towards particular types of task instead. Until recently, a useful way to think about the nature of this task-bias was through the ALM hypothesis—machines can perform 'routine' tasks but not 'non-routine' ones. And, as we saw before, many of these 'non-routine' tasks are not just found in high-skilled work, but lower-skilled or lower-paid work as well. In computer science this is known as Moravec's paradox: that many of the things that human beings find simplest to do with their hands are often the hardest to automate. But what it means is that, in the twenty-first century, the level of education that a human being requires to perform a particular task is less and less informative about whether or not a machine will find it difficult as well.

A final implication is the uncertainty that these recent changes introduce into any attempt to think about the future of work. An earlier speaker mentioned the importance of humility, and I think that is a very important mindset to adopt. One of the benefits of the ALM hypothesis was that it had an attractive conceptual clarity to it: machines could perform 'routine' tasks but not 'non-routine' tasks. That clarity no longer exists. This means that the future of machine capabilities and, in turn, the future of work, is far more ambiguous than many economists might have imagined 10 or 15 years ago.

As noted at the outset, this talk draws explicitly on existing writing and research, including material that I developed with my co-author, Richard Susskind. For example, see the following references.

References

Susskind, D. (2017, November). 3 Myths about the Future of Work (and Why They're Not True), A TED Talk Delivered in London, November 2017.

Susskind, D. (2019). Re-thinking the Capabilities of Technology in Economics. *Economics Bulletin, 39*(1), A30.

Susskind, D. (2020). *A World Without Work.* London: Allen Lane.

Susskind, D., & Susskind, R. (2015/2017). *The Future of the Professions.* Oxford: OUP.

Susskind, D., & Susskind, R. (2018). The Future of the Professions. *Proceedings of the American Philosophical Society, 162*(2).

14

Two Myths About the Future of the Economy

Nick Srnicek

With technology rapidly changing, there is no shortage of prognostications about what the economy of the future will look like. In this paper, I want to critically examine two common, but mythical, images of the future economy, with a particular focus on work and technology.

Myth #1: Uber Is the Business Model of the Future

The first myth I want to tackle is that Uber is the model for the future of the economy. This is the idea that we are going to see an Uberisation of the economy, whereby more and more firms will take on its business model. We can see this in the numerous new apps and platforms that label themselves as an 'Uber for X' in the hopes of having some of Uber's success pass on to them. Taken to the extremes, we even see an Uber for toilets in the form of Airpnp, which allows users to find publicly shared

---

N. Srnicek (✉)
Department of Digital Humanities, King's College London, London, UK
e-mail: nick.srnicek@kcl.ac.uk

© The Author(s) 2020
R. Skidelsky, N. Craig (eds.), *Work in the Future*,
https://doi.org/10.1007/978-3-030-21134-9_14

toilets in private homes. The 'Uberisation' of the economy is also often taken to mean that Uber's peculiar employment relationship with its workers will be replicated and expanded across the entire economy. This belief in an expanding Uberisation is a particularly pernicious myth and I will try to explain why.

First of all, what is Uber's business model? They are what I have elsewhere called a 'lean platform'.[1] They aim to be very asset light: they try to own as little as possible. Uber, for instance does not own the cars; they do not have to pay for fuel; they are not responsible for car insurance or maintenance or anything like that. Even in the core of the business, they do not own massive computer servers or anything. Instead, they rent them out from platforms like Amazon Web Services. Effectively, the Uber model has been to try to own as little as possible. But what they do own is the technological platform that connects riders with passengers, and that is the source of their value extraction.

The problem for Uber (and likeminded companies) is that this model has not been very profitable. Lean platforms in general tend to have very low margins. This works for some services, such as things that are very high frequency. Taxi services are a good example of this as, in a city like London, at any given time, there will be a large number of people who need rides. With high frequency services, even a low margin business can still make a decent profit. The problem is that a lot of services are not high frequency. For instance, grocery shopping only requires people to use the service once every few weeks—with the result being that a number of these companies are struggling to survive. In effect, the Uberisation of low frequency services seems largely a non-starter.

We can also look at the Uberisation of high-skilled jobs, but here we find a lot of problems as well. You can imagine a scenario where, if you have a high-skilled job, you get onto a platform and start making some contacts with customers. However, given that the platform is taking a cut of every service you offer, eventually the best option for you is to leave that platform and go and start an independent business. This is exactly what a lot of companies have found when they try to bring high-skilled workers onto an Uberised platform: the workers end up going independent

---

[1] Srnicek (2016).

since it makes more economic sense for them. One of the more prominent examples of this is Homejoy, which was a sort of Uber for home cleaning. It collapsed after many of its cleaners decided to leave the platform, because they could make more money elsewhere.[2]

Growth in the sharing economy and in these lean platforms tends to be premised not on being profitable right now, but on the promise that at some point in the future they will be profitable. At the current moment, most of these companies are losing significant chunks of money. This is a growth before profit model, which dictates that losses are part of the strategy. Uber is actually the worst offender here. Uber lost $1 billion a year to fight off a Chinese competitor, where it eventually gave up and moved out of China.[3] It lost an estimated $3 billion in 2016, $4.5 billion in 2017, and in 2018 its losses have been accelerating (despite significant revenue growth). It is astonishing that a company can lose $7.5 billion in two years, have never made a profit in its entire existence, and yet still be heralded as the next big thing for capitalism.

Rather than survive by making profits, Uber survives through venture capital welfare: constant injections of new funding from investors. Looking closely at Uber's funding rounds, what becomes apparent is that there is more and more suspicion from the investors. In the most recent funding round, for instance, the investor group SoftBank actually demanded that Uber take a 30 percent cut on their very high valuation.[4] Effectively, Uber is finding it increasingly difficult to convince investors of its ability to generate profits even in the long-term.

Uber also faces future challenges. The first example of these is regulators. The expansion of Uber's particular employment relationship—where workers are deemed contractors rather than employees—has only succeeded by running ahead of regulators and introducing these new labour practices before regulators know what to do. Regulators are now catching up though, with London being a prime example. There have been court cases about the way in which Uber handles its employees. And Uber is presently facing the threat of being banned from London due to

---

[2] Farr (2015).
[3] Jourdan and Ruwitch (2016).
[4] Somerville (2018).

avoidance of regulators' requests.⁵ So regulators are putting significant restrictions on what the Uber model can do.

The other challenge is that Uber and other lean platforms are facing worker struggles. After an initial setback as workers were unsure how to organise and fight for their rights in these new business models, the last year has seen workers striking back in increasingly significant ways. Uber drivers, for instance are attempting to build unions; Deliveroo drivers are attempting to as well; and many of these lean platform companies are facing a number of lawsuits. Uber had to pay $100 million in one settlement; Lyft had to pay $27 million in another settlement; Postmates is currently facing an $800 million suit.⁶ One lawsuit for Uber estimated they would owe drivers $852 million if they were deemed employees and not independent contractors. Uber retorted that it would only be $429 million.⁷ The result of this worker pushback is that these very low margin businesses are going to become even more unprofitable in the future, and the business model is unlikely to expand much further.

What is Uber's plan? Here we see that even Uber doesn't think the business model they pioneered is likely to succeed. They want to grow big—to monopolise taxi services. Yet their next goal is to replace drivers with self-driving cars and build a massive moat around their business that no one else can compete with. This is a major shift in the nature of their business as suddenly they are taking on the costs and responsibilities of an immense amount of fixed capital.

We can see the shift by looking at a now famous quote from 2015:

> Uber, the world's largest taxi company, owns no vehicles. Facebook, the most popular media owner, creates no content. Alibaba, the most valuable retailer, has no inventory. Airbnb, the largest accommodation provider, owns no real estate.⁸

---

⁵ I am less convinced that Uber will ever be banned from London and think that this threat is more of a negotiating tactic than anything else, but it does show the regulators are cracking down on this sort of business model.
⁶ Kosoff (2017).
⁷ Levine and Somerville (2016).
⁸ Goodwin (2015).

Today, the reality is vastly different for all of these companies, and we'd need to rewrite the quote for 2018:

> Uber is buying 24,000 cars, Facebook is spending $1 billion on original TV content, Alibaba is spending $2.6 billion on physical retail, and Airbnb is opening branded apartment buildings.

These companies realise that the standard Uber business model does not work, and they are now moving into a much more traditional business approach. So the first myth—that Uber is the future of the economy, either as a business model or as an employment practice—is nothing more than misplaced hype.

Myth #2: AI's Major Economic Impact will be Through Automation

Myth number two is that artificial intelligence's (AI's) economic impact will be through automation. Most of the media attention, think tank analysis and political rhetoric focus on AI as a threat to jobs. And to the extent that a debate is happening, it is about the extent and speed of that automation process. To me it seems undeniable that AI will automate at least some tasks, and likely many tasks, out of existence. However, this will not be AI's biggest economic impact.

To understand what AI's impact will be though, we first need to understand platforms. What are platforms? Essentially, they are intermediaries and infrastructure. They are intermediaries in that they connect different groups together and their infrastructure enables these groups to interact on these platforms. A classic example is something like Facebook, which connects advertisers, companies, users, content producers, app developers, and so on. Facebook allows these groups to interact on the platform and, crucially, Facebook can then collect data from those interactions and produce value from it.

Notably, this idea of platforms excludes one company that we often think about as a leading tech company: Apple. Apple has some aspects of being a platform, embodied in iTunes and the App Store, but the vast majority of its profit comes from selling luxury consumer goods (iPhones, iPads and all the exorbitantly pricey add-ons they can create). That is where Apple makes most of its money, and that makes them a rather traditional business model.

Platforms are far more interesting, in no small part because they are far more alarming. In the first place, platforms are natural monopolies. They have a monopolistic tendency that emerges not through any sort of artificial means, whether collusion or mergers. Instead, the very nature of a platform business tends towards monopolisation. This is for a few different reasons. One is network effects: the more people who use a platform, the more valuable that platform becomes for everybody else. Again, Facebook is a good example. You may despise Mark Zuckerberg, hate the surveillance practices of Facebook and dislike the way they have handled fake news, but if you are going to join a social media network it will be Facebook, simply because that is where all your friends and family already are. That is the power of network effects, and once they reach a crucial tipping point, they grow and grow and grow.

The other aspect that leads to a monopoly is the ability to extract and control data. By situating themselves between all these different groups, platforms position themselves in a space where they can collect a lot of data. Any interaction that happens on the platform becomes a piece of information that can then be fed into things like machine learning. If data is the new oil, platforms are the new oil rigs. Their intermediary nature allows them to build a moat around their business since as they collect more and more data, it become increasingly difficult for the competitors to beat them. The result is again a tendency towards monopolisation, as the data-rich get richer.

The final reason for the monopoly tendency is path dependency. Once a platform becomes dominant, they create a whole series of entrenched and dependent groups interested in maintaining the platform's dominance. For example, users invest their time and data in a particular platform, and subsequently become dependent on that platform. If you want to leave Facebook for a new social media site, you lose all your friends, your connections, your data, your content, your personalisation and so on. Similar dynamics hold for developers as well, who will often start to tailor their products towards a particular platform (hiring people skilled in programming for it, or marketing on it). The result is that as a platform becomes dominant, many people start to develop an interest in maintaining that dominance.

So for all these reasons, platforms tend to become monopolies—but crucially, at the moment, they are monopolies in individual services. Facebook controls social networking; Google controls search engines; Amazon controls ecommerce. They are dominant in those fields, but those fields are themselves relatively small in terms of the overall economy and even the overall digital world. This siloing of platform monopolies also means they have been able to coexist relatively harmoniously, because they each have their own independent area. There has not been a lot of direct competition between these major platforms. But AI changes all this.

Contemporary AI involves machine learning, which is an approach that involves throwing massive amounts of data at a problem, training algorithms to produce something that can learn patterns and making predictions on the basis of the trained algorithm. This approach to AI requires a lot of data, and therefore companies with access to a lot of data are in a privileged position. As we have seen, platform monopolies are precisely those best placed to take advantage of machine learning technology. Yet to maintain their lead—particularly against each other—these companies face a structural imperative to extract more and more data. This is not only a quantitative increase in a particular type of data (say, geolocation or financial data), but also a qualitative increase in the kinds of data that are being collected.

Partly because of this, what we see is all of these companies starting to expand out from their core business to other places. Amazon is no longer just an ecommerce company; it is getting involved in cloud computing, media content, logistics and the consumer internet of things, to name just a few endeavours. Likewise, with Google and, to a lesser degree, with Facebook. Both are investing and buying up companies all across the tech space in areas that offer new data extraction possibilities. The old monopolies were based on vertical or horizontal integration—but today there is a more rhizomatic integration based upon data as a resource. The end result is that these companies are no longer being siloed into single markets, but instead are becoming general purpose AI companies. And at that level, they start to become competitors.

We are already witnessing emerging competition between these companies over the collection of data. For instance, Google Home versus Amazon Alexa is a key proxy battle in this war, with each of them making efforts to harm the other.[9] There is obviously competition over smartphones; there is competition over personal assistants; and the major new front is now competition over cloud computing. The once peaceful harmony between these platform monopolies is now becoming a significantly more contentious space as they encroach on each other's territory.

This leads me to a core point, which is how AI affects monopolisation tendencies of the platform economy. AI has its own virtuous cycles: more data means better AI, better AI means better services and products, better services and products means more users, and more users means more data. So as a company improves its AI competences, it tends to pull away from other competitors. And as companies with extensive data extraction are the most likely to benefit from the virtuous cycle, it means a further consolidation of power, data and resources in the hands of the few companies that already dominate the platform economy.

This is all the more important because AI is a general purpose technology.[10] AI's impact will be felt across the economy—so those who control AI, who can actually do AI, are going to have major power and influence in the economy. For instance, for economic impacts we might think about Amazon Web Services or Google Cloud building up AI that they can rent out to other businesses that are too data-poor to build their own AI. They will effectively rent out the basic infrastructure of the digital economy. Likewise, we can imagine companies becoming the sole provider of a particular service. Facebook, for example, already uses such dominance in the attention economy to shape otherwise powerful media companies. There are also the political impacts. These companies are already determining who can access these services and under what conditions, and without any civic accountability we would expect of a demo-

---

[9] Wetzel (2018).
[10] Jovanovic and Rousseau (2005).

cratic service. They can also block out competitors, by simply copying them or purchasing them. And even if unintentionally, these companies will wield massive amounts of power over the companies and publics dependent upon them. For instance, when Facebook makes a slight change to its algorithms, numerous media companies see their traffic plummet and their profile decline. It is not difficult to imagine this power being wielded in ways that perpetuate the power of these platforms.

To be sure, AI's impact on the labour market through automation of tasks will very likely be significant. However, the media and scholarly attention paid to this channel of influence have overlooked the far more significant impact emerging from the concentration of power in a handful of global companies.

References

Farr, C. (2015, October 26). Why Homejoy Failed. *Medium*. Retrieved from https://backchannel.com/why-homejoy-failed-bb0ab39d901a#.a4l51hejy

Goodwin, T. (2015, March 3). The Battle is for the Customer Interface. *TechCrunch* (blog). Retrieved from http://social.techcrunch.com/2015/03/03/in-the-age-of-disintermediation-the-battle-is-all-for-the-customer-interface/

Jourdan, A., & Ruwitch, J. (2016, February 18). Uber Losing $1 Billion a Year to Compete in China. *Reuters*. Retrieved from http://www.reuters.com/article/uber-china-idUSKCN0VR1M9

Jovanovic, B., & Rousseau, P. L. (2005). *General Purpose Technologies*. Working Paper (National Bureau of Economic Research, January 2005). Retrieved from http://www.nber.org/papers/w11093

Kosoff, M. (2017, November 9). Why the 'Sharing Economy' Keeps Getting Sued. *Vanity Fair*. Retrieved from https://www.vanityfair.com/news/2017/11/postmates-worker-classification-lawsuit

Levine, D., & Somerville, H. (2016, May 10). Uber Drivers, If Employees, Owed $730 Million More: US Court Papers. *Reuters*. Retrieved from http://www.reuters.com/article/us-uber-tech-drivers-lawsuit-idUSKCN0Y02E8

Somerville, H. (2018, January 19). Softbank is Now Uber's Largest Shareholder as Deal Closes. *Reuters*. Retrieved from https://www.reuters.com/article/us-uber-softbank-tender/uber-softbank-deal-has-closed-making-softbank-largest-shareholder-idUSKBN1F72WL

Srnicek, N. (2016). *Platform Capitalism*. Cambridge: Polity Press.
Wetzel, K. (2018, March 6). As Amazon and Google Keep Bickering, Consumers are the Ones Who Lose. *Digital Trends*. Retrieved from https://www.digitaltrends.com/home/how-far-will-amazon-and-google-take-this-bickering-over-youtube-and-nest/

Part VI

AI, Work and Ethics

15

AI, Ethics, and the Law

Cathy O'Neil

The nature of technology in general is to replace repetitive human work with machines, and the nature of algorithms more specifically is to replace repetitive bureaucratic or rule-based processes with automated decision-making.

That doesn't necessarily mean a given worker is fired because they're replaced by an AI (artificial intelligence). More likely, as the AI gets put in place, it gets trained by the current workers, and subsequently crowds out the need to hire humans later.

This well understood dynamic is important for two reasons. First, because it will affect the availability of jobs in the future, especially in repetitive or rule-based fields and especially if the workers in those fields are well paid or hard to train in developing countries. In this highly speculative essay, however, I'll focus on the second reason the dynamic is important, namely the potential ethical consequences of replacing humans with automated decision-makers in the field of law.

Specifically, I will examine how AI is and will affect the nature of law, a field particularly vulnerable to "AI disruption" given the rule-based

C. O'Neil (✉)
ORCAA, New York, NY, USA

systems that undergird it and the expensive workers that toil within it. I'll argue that AI will be deeply disruptive and potentially highly consequential to the larger public good.

After defining an algorithm and giving a problematic science fiction example, I'll work through three versions of how AI is or might be currently changing law. For each example I'll point out some problematic issues that probably exist for most high stakes, widely used algorithms. Taken as a whole, however, an even bigger problem emerges: AI will undermine the very goals of the field of law. Moreover, I will conclude, solutions to these theoretical existential risks are not apparent.

Basic Definitions and a *Star Trek* Example

An (predictive) algorithm, AI, big data or predictive analytics: they're all referring, at least at the highest level, to the same idea, predicting the future based on the past. More precisely, each algorithm is trying to predict some fixed definition of success based on patterns of what led to success in the past.

This might sound technical, but it's also something we do every day when we get dressed: what outfit will be successful today? First you need to define what you mean by success, which on a given day can be changed (do I want to be comfortable? Professional? Warm and dry?) and next you scan your memories, as well as your closet, for outfits that can optimize to that definition of success.

In the case of a formal algorithm, the definition of success does not waver; it's codified in computer code, and that precise concept of "success," as well as the associated concept of the cost of failure, are embedded in a mathematical object called the objective function. Once the data scientist decides on the objective function, and the historical training data, the ensuing algorithm is largely determined.

Sounds simple, and it sometimes is. But when the output of the algorithm (the prediction itself) is used in a powerful way, a feedback loop is created: the algorithm doesn't just predict the future, it causes the future. For example, if an online loan company uses an algorithm to decide who will pay back a loan ("success"), they'll base their decisions on who gets the loan using that prediction. In particular a person, or a community of persons, deemed unlikely to pay back that loan will systematically be prevented from getting loans, even if the prediction is wrong. And even more importantly, even if that prediction was wrong initially, that prediction ends up being correct if enough predictive algorithms agree that "people like them" don't look likely to pay back loans, and they are systematically shut out of the banking system. In this sense predictions become truth, and correlations become causations.

There are plenty of real examples where this touches quite demonstrably on ethics and the public good, but I'll indulge in a fictional example taken from an episode of *Star Trek: Voyager*'s seventh season, called "Critical Care." The Voyager doctor, which is an AI, runs from a holographic emitter, which is stolen by an alien. The doctor is initially forced to work in a chaotic, under-resourced alien hospital filled to the brim with dying people in desperate need of life-saving medicine, but he eventually gets moved to a higher floor, which is beautifully run and caters to well-off folks getting the sci-fi version of botox treatment with the same medicine that was desperately needed only a few floors below. Turns out the whole system is controlled by another AI, called "the Allocator," which decides which patients get which medicine based on their "social utility." Nobody in particular is in charge, because everybody is constrained and controlled by the objective function that was programmed sometime in the past and then never questioned again.

Assuming perfect predictive accuracy, which is never actually possible except in thought experiments, the ensuing feedback loop confusing causation and correlation is in full force: don't bother keeping that population alive, since they won't be socially useful. And, in turn, they won't be socially useful because they're almost dead and have no chance of getting the medicine they need based on how these predictions get used. In the end, the doctor intervenes on the process by injecting a supervisor with the blood from a sick, lowly ranked patient. The computer gets confused by this unauthorized transfer and confuses the supervisor with the original patient, prompting that supervisor to change the definition of success and save a boatload of lives. In other words, a single ethical voice over rode the machine in an unlikely way (realistically, the supervisor was just a middle manager and wouldn't have had access to the source code).

Keep your eye on the following strains of reasoning in the above example: a machine given too much power, creating its own reality, with nobody in particular in charge. What once likely looked like an efficient-minded solution to all problems becomes a public health nightmare.

Three Examples in Law That Are Each Problematic

Legal analytics is an exploding field, with a LexisNexis Legal Analytics report suggests that 92% of law firms tout the cost savings from AI (Becker and Howard 2018).[1] I'll give three examples of how algorithms are apparently taking over rule-based legal processes once performed by armies of human lawyers. After explaining at a high level how each of them works, we'll consider issues of bias, accuracy, lack of accountability, feedback loops, and gaming.

In the near past, the first job out of law school for many graduates would be to spend their days combing through boxes in a warehouse, looking for a smoking gun for corruption or a conspiracy. This was called discovery, and it's given way to e-discovery, the algorithmic version of sifting through evidence provided from the other side in a lawsuit.

The folders have been replaced with emails and pdf documents, and the human sifting has largely been replaced by sophisticated keyword searches. That is, we have a list of keywords, and then we develop a scoring system on any word that determines the likelihood that it is highly related to one of the keywords. This is called "fuzzy matching," and it's supposed to take care of misspellings, intentional or not, that might end up being smoking guns. For example we'd want to recognize "blackma1l" as pretty close to "blackmail" and if "operation stork" is a keyword phrase, we'd also want to be on the looking for "stork operation."

Problematic consequences of e-discovery are as follows: first, the employment consequence is that such algorithms get rid of the need to hire workers. The reasoning here is that one smart associate, with thoughtful keyword searches, can do the work of a dozen associates in a dusty

[1] https://www.americanbar.org/content/dam/aba/events/professional_responsibility/2018_cpr_meetings/2018conf/materials/session1_ethics_issues/session1_all_materials.authcheckdam.pdf

warehouse. After all, they're dealing with digital tools. We'll soon see why this reasoning might not be sound.

Second, judges might largely defer to e-discovery algorithms because they're intimidated by the terminology, but there are no standards in the field, no due diligence on the consulting companies doing the work, and at least a theoretical risk that the documents turned over to the other side could be done in a biased or hostile manner. For example, the defense might hand over so much data—terabytes or more—that it's difficult to deal with the sheer data handling (to be clear, this is not a new method, but it's easier in the age of avalanches of digital data). Or they might do a poor job of collecting all the texts, messages, emails, and other documents from all the relevant parties, but again the sheer documentation about the documentation is overwhelming (this would get them into deep trouble with the judge if it's discovered). Finally, they might do a bad job (oops!) matching the search keywords with the documents themselves. In other words, bad fuzzy matching, which is something that can be blamed on the e-discovery algorithm itself.

To be clear, the blame could be shared with humans. E-discovery algorithms are also relatively easy to game, simply by misspelling the key words by more than one replaced letter. Or, much more slyly, by avoiding future keyword searches by using very common words at all times. "I did the thing, and put it on my desk" will never be picked up by e-discovery filters because the words are all too common. And leaving a solitary lawyer, even a genius lawyer, to try to find that is worse than asking them to find a needle in a haystack. At the same time, we won't know how many lawyers to assign to this task, especially when we don't know what we haven't yet found, and when the amount of data to sift through it has grown too large to do by hand.

Finally, e-discovery represents an arms race: the abler, richer law firms will have slyer e-discovery engines which can build in subtle advantages. Again, not a new problem, but a new example, coming from technology, of how the legal system getting less responsive, less transparent, and more protective to those already in power and less available to long shot plaintiffs or defendants.

On the whole, the biggest threat posed by e-discovery is that its failures are hard to measure. And over time we tend to trust a system simply because we don't understand how it works and there is very little to no scrutiny of what might be going wrong. The best way to see the problem might just end up being measuring the correlation of money with outcome and seeing it rise over time.

Next example: there are now algorithms that help choose the composition of juries. This could be a good idea, or at least could have positive consequences, since big data and online profiling techniques might be able to summon people to jury duty whose addresses have changed, which is a surprisingly big problem and skews jury pools toward whiter, richer jurists.

But when it comes to the actual jury selection used by both sides in a lawsuit to pick and choose jurors to reject, it could be problematic indeed. The newest techniques are based on demographic profiling methods that were developed in the realm of political polling and targeting and then perfected by the online advertising industry. In particular, it can predict the position and the "persuadability" of different potential jurors on particular issues, and could even suggest lines of argument for the lawyers to pursue based on who is on the jury, just as political micro-targeting ads are increasingly tailored to the psychological triggers of their audience.

This emotional manipulation approach, typically light on facts and heavy on approach, is of course a problem for the principle of justice. And just as in the above example, it absolutely represents an arms race between the two sides, tilting the playing field toward the party that has access to more and better predictive technology rather than being in the right.

Finally, there is an emerging industry of legal analytics that is predicting the chances for civil suits to win. Where e-discovery cut down on cost, this represents a way of optimizing earnings. Algorithms in this domain predict which cases will win, how much it will cost to prepare, and what the settlement might look like. The algorithms are trained on precedent, primarily, which is to say looking at historical decisions in case law and predicting that future decisions will be similar. That's not to say the probability of working has to be better than 50%, just that the expected payoff has to be higher than the expected cost. It is the financial, free market approach to law, and it goes without saying that it makes a future that is predicted to remain consistent with past practice.

An obvious problem with this "civil suit success prediction" system is that false negatives are harder to measure than false positives, an asymmetry that can be potentially brutal. That is to say, law firms will learn their mistakes by taking on cases that don't work out, but they'll have a much harder time figuring out which cases would have won had they taken them on, especially with a novel approach. For that matter, the algorithms will likely become stricter in what they're willing to take on as they learn what didn't work out after all, leading to increasingly conservative choices. The further consequence of this narrow, asymmetrical system is that it, once again, creates its own reality; it will be difficult for an individual or small party to convince a big law firm to take a chance that their computers don't feel optimistic about.

To be clear, individual law firms already assess all of these things manually: chances that a case will win, cost to pursue the case, and eventual settlement. The difference is that the manual process is less precise and more intuitive, which means for example that our guts might smell the winds of change in a given law or a given judge more than any historically trained algorithm would. But more to the point, there won't be people who spend the time sniffing the wind at all.

The Bigger Problem: AI Stagnates Our Ethics

Given the above, and more that I didn't have time and space for (Robot judges? They're already happening in the Netherlands (Howgego 2019, January 8).[2] Crime risk scores for criminal defendants? Used in more than half the US states already. Algorithms that tell police which crimes are "solvable" and which to ignore? Also in use in the UK (Nakad et al. 2015),[3] how will the nature of law itself change?

For example how would AI react to "cases of new impression" (as judges call them)—such as trans gender issues, when there is no prior history of the issue? Will the algorithms decide that no information means there's no point in bringing the case? If legal thinking is allocated

---

[2] https://www.researchgate.net/publication/307626256_Digitally_Produced_Judgements_in_Modern_Court_Proceedings
[3] https://www.newscientist.com/article/2189986-a-uk-police-force-is-dropping-tricky-cases-on-advice-of-an-algorithm/

to AI, will there be any lawyers left with the expertise to address new issues in the future? And if not, will that lead to a black market in law outside of the legal system, that provides alternative justice outcomes?

When we define success narrowly, based on profit or winnability, and when we train on the past, it is tantamount to perpetuating the past, or even exacerbating it, pushing it through a perverted lens and activating an influential and pernicious feedback loop. If we imagine law as a lagged indicator of our ethical sensitivities, this lag will grow, possibly endlessly. We end up with a set of laws that are obscured by automated, opaque decisions, based on an outdated set of ethics and optimized to commercial interest.

Wait, you might say. Won't there be law firms which play the role of the arbitrager? In other words, if we now live in the finance-ified legal age, won't "hedge fund firms" emerge that work around the edges of the "efficient market"? That take on unlikely cases that might just win and pay off big?

I'd argue that such an analogy is unreasonable; the point is, the average law firm will be firmly focused on big money, not big justice, and the analogous "arbitrage opportunity" would be one of lots of justice obtained that "the market" had overlooked, but there's not a lot of reason to think it will have a huge payoff. The point of law, and the reason it doesn't translate well to AI, is that it's a deeply human process, optimized to an ever-changing and evolving messy concept of fairness and justice. That's not a particularly predictable or simplified definition of success.

Conclusion: No Obvious Solution

This is not some anti-trust problem, if you will, that can be solved by breaking up a huge company and encouraging competition. Instead, it's a natural consequence of many individual law firms' incentives to win cases, to manipulate jurors, to save costs, and to get paid. There's no individual bad guy and it's not clear how to avoid this. In particular, if we shut down the five biggest legal analytics firms once a year, the others would expand to provide their services, and those services would be largely working in concert with each other, because they will all have largely similar definitions of success and historical data.

Possibly the only thing we can do, in order to ensure our legal system doesn't reach full automation, is to take our constitutional right to due process very seriously and insist that the algorithms—all of them, including jury selection as well as case selection—are open to scrutiny in various ways (and by that I mean full audits, not just the source code, which is not always a meaningful form of transparency). But most importantly there might be some things we simply refuse to automate. After all, just as the Star Trek Doctor played the role of ethicist (albeit an automated one! The irony does not escape me!), we need humans to oversee our sense of right and wrong as it is embedded in the law.

References

Becker, J., & Howard, B. C. (2018). 4 Ways that Law Firms Benefit from Legal Analytics. *LexisNexis*. Retrieved February 16, 2019, from https://www.americanbar.org/content/dam/aba/events/professional_responsibility/2018_cpr_meetings/2018conf/materials/session1_ethics_issues/session1_all_materials.authcheckdam.pdf

Howgego, J. (2019, January 8). A UK Police Force is Dropping Tricky Cases on Advice of an Algorithm. *New Scientist*. Retrieved February 16, 2019, from https://www.newscientist.com/article/2189986-a-uk-police-force-is-dropping-tricky-cases-on-advice-of-an-algorithm/

Nakad, H., Jongbloed, T., Herik, H., & Salem, A.-B. M. (2015). Digitally Produced Judgements in Modern Court Proceedings. *International Journal for Digital Society*, 6, 1102–1112. https://doi.org/10.20533/ijds.2040.2570.2015.0135.

Part VII

Policy

16

Policy for the Future of Work

David Graeber

It feels a trifle ironic, my being placed in the "policy" section of the conference, because I once wrote a brief, one-paragraph manifesto called, "Against Policy" (Graeber 2004). It has always occurred to me that "policy" and "opinions" form a set, and a rather pernicious one: that is, "opinions" are what you have when you have no power, so your views on what to do have no effect on actual policy; most people have "opinions" because those who make policy don't much care what they think; "policy," conversely, implies some sort of technocratic elite analysing a situation and imposing their solutions on people who have not, on the whole, been allowed to deliberate on the matter themselves, or even, in many cases, been consulted.

So I don't really like the idea of "policy." Still, if we are simply talking about the practical application of some of the ideas we've been discussing, I think I could make a few comments, and pull various strands together. I've been conducting research about work for some time now, and as it happens I just received the galleys yesterday for a book on the subject I

D. Graeber (✉)
Department of Anthropology, London School of Economics, London, UK
e-mail: d.graeber@lse.ac.uk

have been working on for some time. In the light of this research, I'd say there are two things that most immediately jump out at me about the discussion we've been having—and in this, it resembles many discussions that we have been having about work and the future of work.

The first is that no one seems to remark on the profound irrationality of the framework of the discussion. That is to say that there seems to be a general feeling that the rise of the robots is a terrible thing; it will put millions of people out of work, and what are they going to do? It's assumed automation is going to be a problem. It strikes me that if there was any absolute proof that we are living inside a fundamentally crazy economic system it's that the prospect of eliminating most undesirable or dreary forms of work is treated as a problem. Why should that be a problem? For thousands of years, our ancestors dreamed of a society without work, or in which the need to work would be drastically reduced. Finally, we stand at the brink of such a world and suddenly we do not know what to do. We have trapped ourselves in an economic system that makes that a dilemma because we do not know what to do with all the people who are out of work.

This mindset goes back some time. I do not know whether anybody has ever read *Player Piano*, Kurt Vonnegut's very first novel; it is all about when robots replace all the factory workers and they are all sitting around getting drunk, playing pool and being depressed; that is, it's assumed that if not standing on production lines fusing things together, the majority of the population just wouldn't know what else to do with themselves (Vonnegut 1952). I find it telling that Kurt Vonnegut had dropped out of an anthropology programme at the time he wrote that book. Perhaps if he had finished the programme he might have learned that people around the world have often operated on three or four hours of work a day, as Marshall Sahlins was later to point out (Sahlins 1972/2017). Oddly enough such people do not become listless and depressed. They find all sorts of ways to entertain themselves. Lack of work is not an inherent problem.

So how did we get to the position where the elimination of work or the massive reduction of work is considered a problem? How is it we can't even conceive of an economic system that would, faced with the problem of less demand for labour and more abundance, can't just redistribute the work in a more or less equitable fashion so we can use our free time to enjoy the abundance? This shouldn't be a hard problem! Of all the economic problems one could be facing it's hard to imagine an easier or more desirable one. Yet we're flummoxed by it. We act as if market capitalism by its nature couldn't handle this. Which is odd because they also tell us we have to accept market capitalism as opposed to any other conceivable system because of its amazing efficiency. Suddenly it turns out that in the face of twenty-first century problems, at least, it's completely inefficient.

The second thing that nobody really remarked upon is that this—that the crisis of the rise of robots and the fear of automation has happened before. It happened in the 1930s, but then, right at the end of the 1960s there was another enormous moral panic. I know one person (Win McCormack) who was taking part in think tanks at the time, and he told me that all the Ivy League schools in America were organizing, "what are we going to do when all the jobs are gone, and the working class is thrown out of work." The *Player Piano* scenario felt quite imminent at that time. Then around 1971 or 1972 you get things like *Future Shock* by Alvin Toffler coming out which gives public voice to all this; Toffler makes an argument about what he calls "accelerative thrust," that the speed at which technological change is happening is geometrical: the number of new patents, energy use, and so forth (Toffler 1970). For instance, if you look at the speed at which the fastest person can travel, for example at that time, and it did seem to be increasing at such a rate that it was reasonable to assume that by now, we should be exploring other solar systems. It's a bit ironic that he used the term "accelerative thrust" though because in fact that particular indicator hit its high water mark just around the time he was writing the book, then abruptly stopped: the fastest speed a person has ever achieved was achieved in 1969, with Apollo 10, and we have never gone faster since. Most of his trends started slowing down at just that moment.

Nonetheless, there was a general moral panic at the time, and a lot of it took the form of a very conservative fear of the social consequences of too much wealth, leisure, and rapid technological advance. (It's not insignificant that Toffler himself became a darling of the neocons.) Much of it was explicitly anti-feminist: "What is going to happen to the patriarchal family and when we are all test tube babies?" (People were anticipating Shulamith Firestone long before she wrote.) "What is going to happen when all the working class gets thrown out of work and everybody becomes a hippy?" Obviously this was in the context of the times when it was assumed that there would be efficient welfare states which would redistribute the goods at least to a reasonable degree. One policy result, which can be observed around that time, was a vast shift of research and development away from the "space age" and futuristic technologies popular at the time and towards medical, information, and military technologies—that is, largely to things that were useful for social control. One could make the argument they also started working to reign in the welfare state around that time: anyway, that's what eventually started happening.

Somehow we are at that moment of moral panic again, but this time, with somewhat different ground rules.

As I mentioned, the idea that machines are going to throw us all out of work and that this will be a disaster goes back well before the 1960s or even before Vonnegut; it harkens back at least to the Depression; even arguably to the Victorian age. Keynes coined the phrase "technological unemployment" in the 1930s as one of the main causes of the mass unemployment of the time. As a result, some argue why are we worried now, the structural employment always predicted in the 1930s never happened. Or in the 1940s or 1950s. John F. Kennedy convoked a whole conference on what to do about the imminent unemployment with automation and the eventual emergence of robots—it didn't happen then either, so there's no reason to think this time is any different.

However, an argument could be made that the mass employment predicted since the 1930s actually did become structure—we're just unable to see it. At least this is what I want to propose here. If you look at the kind of jobs that were considered necessary in the times that Keynes was writing—and we were 10 years away from the time Keynes was predicting that we should have a 15-hour week—many, if not most, of the jobs were indeed eliminated (Keynes 1930). Technological unemployment did happen. We could be living just as he predicted. But instead we made up new forms of employment to keep people busy which were, we might say, only made necessary by each other. There's no real objective reason why most of them should have to exist.

Now, we're probably not entirely unfamiliar with this sort of argument but I'm not going to make the one you usually hear. The typical narrative is that we denied ourselves utopia because of the endless creation of new needs: a classic Christian trope, by the way, Fallen Man is cursed by insatiable desires which thus blind him to the dictates of his own reason. The slightly less theological way in which it's usually put is that given the choice between more leisure and more consumer goods, people collectively opted for the latter. We chose consumerism. This narrative of course goes along with the discourse of the rise of the service economy we've been hearing since at least the 1980s, but a lot of that is really just hot air. If you look at the numbers it just doesn't wash. A key question is how you define service work. If you define it simply as it was defined in Keynes' time, as giving people haircuts or serving them coffee, well, you find the number of people employed in services has remained pretty much flat at 20% in most industrialized countries for the last 100 years. There's been changes in composition—fewer domestic servants, obviously, more baristas—but the total numbers have barely altered. What has happened is that information technologies have skyrocketed. Administrative, clerical, managerial, and supervisory jobs have skyrocketed. At the same time farming, largely, and industry declined (though not nearly as much as people say). So what's basically replaced the old factory and farming jobs is not service, per se, but office work.

This whole phenomenon became an interest of mine after I wrote a little essay, which was kind of a thought experiment, called "On the phenomenon of bullshit jobs" (Graeber 2013). I had a friend who was starting a new magazine, and asked me for something provocative. Well at the time I had a kind of list of essays I always wanted to write that nobody would normally publish so I trundled one out. The original essay was really a reflection on the puzzlement I'd often feel when I would meet people at academic parties or spouses of colleagues; I'd ask them what they do for a living, and quite frequently, the result was embarrassment. They said, "Oh, nothing really," or "well, to be honest, not much. I really just work two or three hours a day. Don't tell my boss but most days I mostly just play around on Facebook." I kept meeting these people. Or others would write off their entire line of work, "Well, I am a corporate lawyer, but to be honest, the whole industry is pointless, I kind of wish it didn't exist." So I started thinking: how many people are there like that? And what must the moral and psychological effects be! Imagine waking up every morning, going to work, and secretly believing your job is completely pointless and should not exist? Or knowing that you are just going to pretend to work for the next eight hours? That actually rather got to me because coming from a working class background, as I do, I know that the most awful part of any real job is that part of it is that you have to pretend to work even though you've finished the job, because the boss is looking and you're on the clock so he doesn't want to see you slouching around whether or not there's anything that needs doing. And I thought, "good lord! what if your entire job is like that? What would that be like? Is that what middle class people do all day? No wonder so many of them seem so depressed and empty."

I wrote this little piece saying: maybe this is the reason we do not have the 15-hour week. Somehow, we have conspired to give ourselves these made-up jobs just because we feel that everybody should be working. That there is this incredible moral imperative. It was a thought experiment, but, if there was ever an experiment that was confirmed by the reaction, this would have to be it; because within three weeks of publication—and this was in an obscure periodical mind you, *STRIKE!* magazine, which had recently spun off from *International Times*, an anarchist magazine which hadn't even existed a few months before—well, within weeks, the essay had already been translated into a dozen languages. The server kept crashing. It received millions of hits. I started getting emails from people saying, "I work in financial services. This is so true. I got this essay eight times just today across my desk," which if nothing else shows that many people in financial services really do not have much to do. So the essay, "On the Phenomenon of Bullshit Jobs," started circulating everywhere. People started writing confessionals. There were countless blogs: I think I saved about a hundred of them. People were writing things like, "Yes, it is true. I am a corporate lawyer. I contribute nothing to society. I am miserable all the time" or confessing anonymously online that they could not admit to friends and families what they really do all day, which was, very frequently, absolutely nothing.

So clearly I had identified a kind of taboo, a social issue that simply couldn't be publicly addressed as such. Think about it. Newspaper columnists or TV pundits are the social equivalent of preachers in this day and age, and they're always going on about how young people, and so on, are lazy and workshy; the solution to every social problem always seems to involve more work; can you even imagine such a person getting up and writing a column about how actually a lot of the work we do is pointless and we all need to slow down and relax a little? Work is considered a value in itself, "hard-working" means "deserving," if you don't work hard you're undeserving, and all this is simply hard-wired into our political discourse. So the issue was literally unspeakable.

Yet it clearly was one of massive importance. At one point YouGov did a poll, directly inspired by the essay, I think in 2015, and there was another in Holland a year later. YouGov found that in the UK, 37% of all people who had jobs said that if their job did not exist it would make no difference whatsoever—which is just astounding (Dahlgreen 2015). I'd myself thought the number would be half that—15%, maybe 20% max. In Holland the number was 40%. Only 50% in the UK were absolutely sure their job served any social purpose at all.

In a way this is something we've kind of known for a long time, that a lot of people think their jobs are a complete waste of time; what I'm really proposing here is something that shouldn't be very radical, but apparently is. I'm saying: what if they are right? "Let us assume that these people know what they are talking about." After all who else would know better? If you think your job is useful in some way, I will take your word for it. If you think your job is completely pointless, then I will take your word for it too. But think about the implications. Because there are so many people who would never say their jobs are pointless. If you're a nurse, a bus driver, an exterminator, a grocer… You might not like your job but you definitely know that the work needs doing. And my own research has made it clear that real service work, store clerks, or waitresses and the like, feel the same way. So if 37% nonetheless feel their jobs are pointless, then that means that almost anybody sitting there in an office who you might suspect is secretly thinking "nobody really needs to be doing this" probably is, indeed, thinking exactly that.

Then you have to think about all the support work. If 37% to 40% of jobs are doing nothing then how many people who are cleaners, who water the plants in that building, or work in security—people who are doing real work—are doing their real work so that other people can sit around doing nothing? Then if you include the bullshitization of real work (paperwork and meetings deemed useless by those who do it), which according to some surveys is extremely high, you're definitely talking about over 50% of the work being done in our society being completely unnecessary. Think about that. We could easily institute a 20-hour week. (Obviously, the question would be how, which is where it gets to policy. I will get to that in a moment.)

So: how did we get to this ridiculous situation? It is something of a mystery. I explore a variety of possible answers in the book. One thing we can say for sure: one of the only things the left and the right seem to agree on policy issues is that the solution to any problem is more jobs. And this demand for jobs is somewhat indiscriminate. At least, you never hear anyone say, "We demand more jobs, but only ones that actually do something." Neither do you hear anyone object to policies designed to lower unemployment that some jobs are not worth having. In the same way, when in America or the UK they talk about rich people as "job creators," and thus justify using the tax system to reallocate even more of the national wealth to them, so they can create jobs, no one really says, "oh yes, and make sure those jobs are useful in some way." It is assumed that the market would never produce a useless job, and somehow giving money to rich people and putting political pressure on them to hire people is "the free market," so even if the people doing the jobs feel their jobs are useless, they must be wrong, jobs are useful by definition. At least in the private sector. (Which is another common misconception: if you look at the numbers, bullshit jobs seem to occur roughly equally in the public and private sectors.)

You could say there are at least two levels of causality we need to look at: on the one hand, the internal institutional dynamics of large organizations which tend to create and maintain such pointless positions—and there's definitely an already-existing sociological and even economic literature on this—and the larger moral and political question of why no one does anything about it, or even in some cases, encourages it. I actually found a smoking gun interview with Obama where he actually admitted it: "Sure," he said, "Having a national health-type system or a single payer insurer would be much more efficient. People argue therefore we should have one all the time, but what are we going to do with the office workers? There are two or three billion people who work in the private healthcare industry. If we have an efficient system all of these guys will be out of work" (Sirota 2006). So here we are with the President of the United States saying that a socialist system would be more efficient than a market system but therefore, that he prefers a market system because it will keep lots of people in unnecessary jobs. There is a political will to keep things like this, a recognition on the part of authorities that they want to keep this engine of creating unnecessary jobs going—because after all, you can fire factory workers, or drivers, and tell them it's their own fault, but office workers, that's the core constituency of the democratic party and you can't completely alienate those guys. On the other hand, genuinely changing the system, creating not only socialized health but a more equitable distribution of wealth and labour, well, as far as Obama is concerned that's completely off the table. "Hope" and "change" don't cover hoping for changes like that. This is why I say that in the final analysis, Obama was a conservative. But the result is millions of people as he says toiling away at jobs they know to be socially useless, or worse, and the human toll of that is enormous.

There are another couple of points that I think are really important to make here. One is about the effects of all this useless work on perceptions of value. Historically, it's important to remember that the labour theory of value was almost universally accepted by popular classes in the nineteenth century, particularly in America; there was this incredible outpouring of hatred towards corporate capitalists—"robber barons" as they called them at the time—when they first appeared; and this was followed by an explicit intellectual counteroffensive from the side of the robber barons themselves; starting in America with people like Andrew Carnegie. It took explicit aim at the idea that workers create wealth, or that one's work should be one's primary means of expression, self-realization, or the basis of one's feelings of self-worth. This was startlingly effective. After all, if you said "wealth creator" in 1850, everyone would assume you were referring to workers; if you say "wealth creator" now, they'll assume you mean bosses. This was accompanied by the idea that people should think of themselves as valuable according to what they consumed instead. The obvious problem here is: how do you validate labour in a situation like that? Other than simply as a means to earn your consumer toys since that didn't really cut it, in moral terms (and remember, the US is a very moralistic society.) More and more, the answer was to fall back on the old puritan principle that work is of moral value in itself.

If you flip through the sociology of work literature, or surveys about work satisfaction in rich countries, you almost always find yourself face-to-face with the same paradox. On the one hand, (a) people find their sense of self-worth and being in the world from their work; on the other, (b) most people hate their jobs. It's very hard to imagine how people could think both these things are the same time but clearly many people do.

The tradition of Puritanism—which by the way goes back much further than Calvinism, to Medieval or even some early Christian ideas—provides an answer. People feel validated, they get their sense of self-worth from their work, because they hate their jobs. Work is a kind of secular hair shirt. It is supposed to be miserable. It is this suffering which provides the spiritual legitimacy which justifies the comforts and pleasures of consumption. The result is a feeling that the more pleasure and fulfilment you get out of the work the less legitimate it is, the less it's really work, certainly, the less you should be paid for it. (Everyone feels this way. How many of us who, say, do something that's actually interesting for a living, that they enjoy, haven't caught themselves thinking "I can't believe I get paid to do this!" This is even in cases where the work is providing an obvious social benefit, like advancing science, or providing entertainment.) There is a very deep moral perversity in these feelings, which cause us to feel that jobs that are satisfying should not be as highly paid as those that make us miserable. It would make sense if it was compensation for taking on unpleasant or dangerous jobs like, say, sewer maintenance, firefighting, or industrial fishing. But in fact these are often poorly paid as well: partly, because they are so necessary. It as if even the satisfaction that comes of knowing one is actually doing something useful for other human beings, that one is improving the world in some way, counts against the misery-value of the work, and therefore justifies worse conditions, less pay, and less overall social respect. There are always a few exceptions to every rule, but generally speaking the result is an overall negative correlation between social utility and pay. Jobs that are obviously useful do tend to get paid much less than jobs that largely are not, or anyway those who have those jobs feel are not.

What I find fascinating—and not a little bit disturbing—is the fact that so many people have come to feel that such arrangements are morally right. "You wouldn't want teachers to be paid too much, because you wouldn't want people who are just in it for the money taking care of our children." People say things like that all the time. (Oddly enough you never hear anyone say "You wouldn't want bankers to be paid too much, because you wouldn't want people who are just in it for the money taking care of our money"—which one might think was a much more obvious danger, but we'll leave that aside for the moment.) It causes this fascinating political resentment whereby—in America you see this all the time—right-wing activists are able to whip up resentment against teachers, effectively saying: "You are supposed to be self-sacrificing. And you get the pleasure of knowing you benefit our children! How dare you want pensions, vacations, good job security and tolerable work conditions too!." Even auto-workers, "you get to make cars, shouldn't that be enough for you? And you expect to be paid $28.00 an hour just because you're providing people with something they actually want?" Similarly, in this country, you see the same weird moral kink in the resentment against people working in the public services; after the financial crash, there was a rhetoric of common sacrifice, but everyone was willing to accept that the bankers who caused the crisis didn't need to make significant sacrifices, aside from a little public shaming, and all those legions of pointless office flunkies didn't have to make sacrifices, they had to suffer already in their knowledge of their own parasitism; but they did demand sacrifices from ambulance drivers, nurses, or firefighters. There is a sense that those people are supposed to be self-sacrificing. Why else had they chosen low-paying, or relatively low-paying, but vitally useful lines and therefore high-minded work? They are doing good in the world; now they can do some more by taking a pay cut. People might not have been delighted that the bankers got off free, but the political party that proposed these policies did get re-elected. Arguably, twice! Certainly they weren't considered monsters and unceremoniously booted out.

There is a perverse inversion of values here, but it's a direct result of this notion of work as a form of self-sacrifice, self-discipline, and self-abnegation.

As for the more mechanical question of the internal workplace dynamics that lead to the gradual accretion of such jobs, this is interesting but I probably do not have time to go into it in any detail. But it's clear that the financialization of the economy has accelerated tendencies that already existed in any large organization, and often quite rapidly. I didn't do quantitative research on the topic, but I did do some qualitative research and some of it was quite revealing in this regard. I solicited testimonies on social media, set up an email account, and received over 250 testimonies ranging from one paragraph to 18 pages in some cases—whole strings of bullshit jobs one after the other—and then followed up with the more revealing ones with often quite detailed questioning.

One of the more interesting testimonies was from an efficiency expert at a series of banks. He was technically a security expert, but his job was to study internal operations, then suggest reforms that would both streamline operations and make them more secure. He said that in his own estimation—and I guess no one would be in a better position to know—80% of people who work in the average large bank were completely unnecessary, either they were doing nothing, or they could easily be replaced by machines. Most of them, he added, were not aware of the supernumerary nature of their jobs: everything was organized in such a way that no one really understood the larger processes they were part of, so they just assumed those processes were not completely absurd. He also said that in 15 years, no reform he'd proposed was ever adopted. Every time he proposed a plan to get rid of some of this waste, it was eventually shut down because it would always mean that some executive would lose out on the number of people they had working under them, and this would be a major blow to their standing. You see, one's prestige within a large corporation (often, even, one's pay) is based upon how many underlings you have, and when someone realizes, "wait, this means I'm going to lose 25 of them," panic ensues. So his every suggestion was vetoed by someone higher up, until he finally realized he had a bullshit job because he was just there to make the bank look like it had an efficiency programme when in fact it didn't (Graeber 2018).

Another surprising thing I learned was that financial firms—basically, large operations in the Finance, Insurance & Real Estate (FIRE) section like banks, accountancy firms or insurance firms—whose business centres on distributing large amounts of money, will often intentionally mistrain people or otherwise take measures to ensure maximum inefficiency. I got one testimony from someone who worked for one of the big five accountancy firms that was handling Payment Protection Insurance (PPI) distributions, who said the company was intentionally training people wrong, putting offices in the wrong cities, destroying documents so they had to be created again, all because they knew that longer it took to distribute the money, the more of it they kept. It is a little bit like Jarndyce and Jarndyce in *Bleak House*. You want to have layers and layers of unnecessary bureaucracy if you are running a basically top-down redistribution of the economy, rather than one that's primarily organized around industrial production; the more you have financialization the more this kind of inefficiency pays. And the logic that starts in the financial sector slowly becomes the norm and extends everywhere.

It definitely extends to universities. This is my riposte, incidentally, to *The Economist* who wrote a reply to the original Bullshit Jobs article almost instantly after I wrote it. They tried to make the argument that this endless creation of new office jobs is actually necessary—it's all because with complex global supply chains, production has become so digitized and efficient that we need many times more people to manage it. So bullshit jobs they claimed were the equivalent of the boring alienating factory job of the 1940s or 1950s, but they are also equally necessary. Our wealth depends on them.

To which the obvious reply is: well then why is it happening at universities? What's the academic equivalent of global supply chains, containerized shipping, Japanese style "just in time" production quotas? It is not like education or teaching at universities has become all that more complicated than it was 50 years ago. We are basically doing the same thing. But somehow, all of a sudden we need three times as many people to administer us while we're doing it. How did that happen? If you look at how it happened, it is quite clear. The number of administrators has gone up slightly in relation to both speakers and students, but the number of administrative staff has almost tripled.

What's more, in America, where it is possible to compare public and private universities, we find the rapid growth of administration is happening faster in private institutions than in public ones. Overall, numbers have tripled. Why? It is largely because every big shot administrator they hire now, every Vice Provost or Strategic Dean, has to feel like they're a corporate executive and that means not only a six-figure salary but that you're automatically assigned two or three flunkies when they come in—because after all, you're not a real executive unless you have two or three subordinates. They hire the assistants first. Then they figure out something for those assistants to do. So what do they do? Generally, they make up new forms of paperwork for people like me to do, time allocation studies, learning outcome summaries, elaborate reports justifying departments to continue to receive the same funding they already are. These kinds of dynamics exist everywhere. I call the results managerial feudalism. You can see the same thing in most large corporations. Layers and layers of managers are added and in between the producers and the top of the system, and the process reproduces itself in every field, starting from finance and large bureaucratic corporations but gradually becoming the model even for the creative industries: so that you have curators in art; producers in addition to editors in the news; in movies and TV writers now complain there are often five, six, even seven layers of suits in between you and the Executive Producer, and every single one of them feels they have to weigh in and change something. All of them tinker with the script and the results are mush.

This kind of feudalization, with its hierarchies of managers and sub-managers and sub-sub-managers, has infected all types of organizations, public and private.

The question I've been asked here is: what are the "policy implications?"

It's pretty obvious you can't approach a problem like this head-on. In British academia we talk about the "creating committees to discuss the problem of too many committees problem." Try to set up a government initiative to address the problem of bullshit jobs and it'll just be the same thing: they'll end up creating more of them. A viable solution would have to go deeper, to question our assumptions.

For instance, in all the discussions we have been having today, every intervention has simply taken it for granted that jobs are necessary, that if a job exists, there must be a good reason for it. There seems strong reason to believe this isn't true. What if we instead started our policy discussions with the assumption that a lot of jobs are not necessary, and that the people who have those jobs know they are not necessary and are simply not in a position that they feel they can speak about such matters because the alternative would be to be thrown on the tender mercies of the unemployment system?

This is why I think the plague of bullshit jobs, and the misery it causes, is one of the best arguments we could make for universal basic income. One of the odd things about universal basic income is that it's backed by such a broad spectrum of economic and political thinkers, from Martin Luther King to Milton Friedman, but this is partly because different advocates are actually advocating quite different things. One might say there's three broad versions of basic income. There's the liberal version, where you are basically giving everyone an income supplement, that's nonetheless calculated to be not quite enough to live on. I think Obama endorses this now. That makes sense: "progressives" or left centrists, liberals, nowadays are basically conservatives insofar as they're mainly interested in conserving the system in more or less its current form. Then there's a right-wing version, which is basically about using a guaranteed income to lower the domain of unconditionality in other parts of the welfare state, or what remains of it: health, education, or housing. That is what people like Milton Friedman were endorsing.

But there's also a left-wing version, which is about entirely severing livelihood from work—which means radically expanding the domain of unconditionality (since one would leave free health, education, etc. intact—and it would probably also require a degree of intervention in the housing market to prevent rentiers from gobbling too much of it up.) In this radical version you give every individual an income adequate to a rudimentary but comfortable life and then let people decide for themselves what they want to do with themselves, how they want to contribute to society. One might refer to this situation as "economic freedom." It sounds strange to us because we've come to identify economic freedom with the right to sell ourselves, or at best to own a piece in our own collective enslavement, but for most people in history of course freedom meant the right not to sell oneself, or to otherwise be reduced to working at another's orders. The normal reaction when you propose something like Universal Basic Income (UBI) of course is precisely, fear of freedom: "But if you just leave it up to every individual to decide what they want to do with themselves, they'll simply lounge around and not work," because they're lazy and won't be able to figure out what to do with themselves, "or, how do you know they won't do something stupid?"

This is incidentally why those who hail from the professional-managerial classes, but nonetheless recognize radical measures of some sort are required, often prefer a job guarantee (JG). But historically, such programmes always create even more bullshit jobs. The only way to ensure that it wouldn't, would be to create a job guarantee on top of UBI, so no one would be forced to take a pointless job just to keep a roof over their heads; in which case, JG would just be a way of providing help finding useful work until people who now have time on their hands begin to self-organize enough they don't need bureaucrats to do that for them anymore. But certainly as an alternative JG would be disastrous.

Also it's based on false premises. First of all, it's perfectly clear that people do want to do something with their lives, so it's not like very many people given UBI would just sit around doing nothing all day; economics teaches us people want something for nothing, or for the minimum output possible, but if that were really true, people paid handsomely to do nothing all day would be happy as clams and in fact they almost invariably report themselves miserable. So then the next line of objection is "sure, but if you just let everybody contribute to society in any way they want, half of them will decide they're poets or try to invent perpetual motion devices, you'll have all these annoying street musicians and mimes wondering around, there will be crime and drug addiction, civilization as we know it will begin to disintegrate." This is what I mean by fear of freedom. Just take the point about useless jobs. How many bad poets are we really going to get? About 2%, 3%, 4% of the population? Obviously not even that. Meanwhile, right this very moment now we are in a situation where 37% to 40% of people in jobs already think their jobs are completely useless and, not only that, not even fun. If you are up there highlighting a medical form all day so someone can get a tax cut, or bribe a politician to insert their accountancy firm in between two health providers, you are not having fun. But no one would ever do a job like that except for the money. I can still imagine people doing sewer maintenance, or removing landmines, or becoming morticians or joining the merchant marine even though they didn't have to, especially if they got additional money for it (which people would provide as these jobs are actually necessary), but no one is going to be doing paperwork that serves no purpose. So if all those people quit and form jug bands, even if those bands are not very good, or even start researching alien abductions or decide to set the world record for having sex at an advanced age, well, they'll be a lot happier, it's very hard to imagine over two thirds of them will come up with something everyone else considers entirely useless, and of course, if one of those poets does turn out to be a Shakespeare, one of those musicians does turn out to be a Miles Davis, one of those crank scientists actually does invent a teleportation device or warp drive, society will benefit more than we can count.

I would make a radical suggestion: that technological unemployment has already happened, that we are in a state of collective denial, effectively, we have decided that rather than opt for collective liberation we're effectively torturing each other out of sheer resentment at the idea someone else might be getting off easy without having to work. But the means for creating a sane society exists. It might seem radical, but it would be easier and beneficial because much of the work of automation has already been done. We just need to lift the veil on what is really going on around us.

References

Dahlgreen, W. (2015). 37% of British Workers Think Their Jobs are Meaningless. *YouGov*. Retrieved August 2, 2019, from https://yougov.co.uk/topics/lifestyle/articles-reports/2015/08/12/british-jobs-meaningless

Graeber, D. (2004). *Fragments of an Anarchist Anthropology*. Chicago: Prickly Paradigm Press.

Graeber, D. (2013). On the Phenomenon of Bullshit Jobs: A Work Rant. *Strike! Magazine*. Retrieved August 2, 2019, from https://strikemag.org/bullshit-jobs/

Graeber, D. (2018). *Bullshit Jobs*. London: Allen Lane.

Keynes, J. M. (1930). Economic Possibilities for Our Grandchildren (1930). In *Essays in Persuasion*. New York: Harcourt Brace.

Sahlins, M. (1972/2017). *Stone Age Economics*. Reprint. Abingdon, Oxon and New York: Routledge Classics.

Sirota, D. (2006, June 26). Mr Obama Goes to Washington. *Nation*.

Toffler, A. (1970). *Future Shock*. New York: Random House.

Vonnegut, K. (1952). *Player Piano*. New York: Delacorte Press.

17

Automation and Working Time in the UK

Rachel Kay

Why Discuss Working Time?

Working time has recently arisen as a contentious topic in the UK media, particularly following the Trades Union Congress' proposal that we should aim to instate a four-day week by the end of the century. Before asking whether this is either desirable or feasible, however, I would like to discuss why working time has become an object of analysis and contestation in the first place. This requires us to make a short digression into the history of norms around work, providing necessary context to the discussion by helping us to understand why particular norms around working time exist today. A historical perspective reveals that today's norms are contingent and not set in stone. Crucially, this prompts us to recognise that change in our organisation of working time is a possibility.

The rise of factory work during the industrial revolution brought about a widespread scrutiny of working time. This era saw a greater centralisation of paid employment in work*places* and the concomitant determination of conditions of work by an employer. This was opposed to the prior

R. Kay (✉)
Centre for Global Studies, London, UK

norm in the manufacturing industry of making goods from one's home. The widespread movement from the home to the centralised workplace is perhaps the biggest change to have occurred in the history of work: for most of human history, work was closely integrated with, or even indistinguishable from, the rest of life. With the centralisation of manufacturing work came a shift from being paid per item produced to being paid per hour spent in the workplace. As a result, working time became popularly contested. This demarcation of working time may be one reason why, whilst we attach a diversity of meanings to the term 'work', a dominant usage today is with reference to paid employment: when we speak about 'working time', this is what we mean, rather than work simply as purposeful or 'instrumental' activity (Volf 1991). This alternative understanding of work certainly deserves attention: I will return to it later on.

The nineteenth century therefore saw the growth of a labour movement demanding shorter working hours and the introduction of legislation relating to working time. The labour movement made significant progress. In the industrialised countries, the average working week was roughly 60–70 hours in 1870. Since then, reduction in working time has been more or less continuous, with the most rapid decrease from 1900–1940. This was brought about by a combination of legislation on working hours and holidays, trade union pressure for a shorter working week and business initiative. This reduction in working hours was so marked that in 1930 John Maynard Keynes made the famous prediction that in around 100 years' time we would all be working a 15-hour week. Keynes argued that with technological progress, labour productivity would rise, meaning that people would have to work increasingly less in order to satisfy their needs.

However, Keynes' prediction was not realised. During the post-war era, UK weekly hours of work fell less rapidly, and from the 1980s onwards they almost flatlined. Why is it that weekly working hours have not continued to fall? A number of possible reasons have been proposed, including an overall decline in productivity growth from the 1980s onwards, a UK company culture that encourages long hours and a seemingly insatiable desire for consumption goods, inflamed by advertising (Skidelsky and Skidelsky 2013). In my view, changes in institutions have been the most important determinant of working hours, since institutions create and perpetuate norms. This viewpoint echoes the assertion recently made by the International Labour Organisation's Global Commission on the Future of Work that investment into labour institutions should form an essential part of any agenda aiming to ensure decent work (2019). The most important institutions in this case are the trade unions. In the UK, the weakening of unions since the 1980s has reduced worker bargaining power and altered the power relationship between employer and employee. Comparing countries across Europe, I demonstrate that different institutional landscapes have brought about different working norms. This then leads me to the conclusion that in order to change working norms, we should focus primarily upon changing institutions.

The Trades Union Congress has recently found that eight in ten UK workers would like to reduce working time in the future. It has frequently been observed of late that UK employees work more hours, but are less productive, than workers in a number of other EU countries such as Germany, Denmark, Holland and France. As of 2016, full-time UK employees work the longest average weekly hours of all EU countries, at 42.3 (Smith 2018).[1] Denmark ranks lowest at 37.8. At the same time, however, around 8 per cent of UK workers want to work more hours than their employers will give them (Stirling and Lawrence 2018). Working too few hours is particularly an issue for the low-paid, who in 2017 worked a median of 27.5 hours per week (D'Arcy 2018). In the UK today, underwork and poverty sit alongside overwork and overconsumption: this is both a social concern and an environmental one. Evidently, working time allocation in the UK is an issue that needs to be addressed.

Arguments in Favour of Working Time Reduction

Since the industrial revolution, two main objectives have motivated the movement for working time reduction (WTR). The first is employee wellbeing. This encompasses both health and safety and WTR as a form

---

[1] As noted by Smith, when full-time and part-time workers are considered together, the UK ranks much lower down the list.

of social progress. For example, the US labour movement's 'eight-hour anthem' of the 1880s demanded 'eight hours of work, eight hours for rest, eight hours for what we will [i.e. leisure]'. Historically, labour unions have typically sought WTR without loss in pay, in order to avoid cutting the pay of low-income earners. The second motive for WTR is to reduce or avoid unemployment by improving the distribution of available work. This has been a particularly strong motive over the past century during economic recessions: for example, following the 2008 crisis some UK employers cut basic hours or reduced overtime (van Wanrooy et al. 2014: 6). In these situations, pay has been cut according to hours lost, but this has been viewed as preferable to outright unemployment.

In the recent debate about working time over the last couple of decades, a number of additional arguments in favour of WTR have come to the fore. WTR is thought to improve productivity of workers due to reduced mental and/or physical fatigue. Organisation for Economic Cooperation and Development (OECD) data has shown that there is a strong negative correlation between annual hours worked and GDP per hour worked across OECD countries (The Economist 2013). This does not mean that there is causation in either direction. But it does mean that decreasing working hours does not necessarily lead to reduced output.

Another set of arguments revolve around gender equality and reducing work-family conflict. Advocates of gender equality have argued for worktime norms that enable a more equitable distribution of paid and unpaid labour between women and men. WTR would free up more time for the unpaid work of childcare and other care-work. This would allow women to participate as equals in the labour market and encourage men to play a greater role in family life at home. Finally, commentators have suggested that WTR will create a more sustainable economy. It is indisputable that society's overproduction and overconsumption is having detrimental effects on the environment, and that this is probably the most urgent problem of our time. Increased labour productivity, rather than being used to fuel greater output, can be channelled instead towards the non-material benefits of increased leisure time. It is clear, however, that the various motives for reducing working time are not all complementary. For example, if WTR leads to increased productivity in some occupations, this is likely to mean that there is no need to hire new workers, and therefore that there will be no alleviation of unemployment.

My view is that the arguments relating to wellbeing, increased motivation at work and work-life balance are the most persuasive. A recent trial of a four-day week to have received significant media attention was carried out at the trust fund Perpetual Guardian in New Zealand. The company decided to make the change permanent after the trial resulted in workers reporting that the above aspects had improved, as well as the workers' productivity increasing. The unemployment argument is less convincing to me, since there is little and disputed empirical evidence to suggest that legislating for WTR leads to increased employment (c.f. Bosch 2000): WTR has only proven useful in times of economic recession, where it simply prevents an increase in unemployment.

Work as Burden or Blessing?

At this point, however, we must ask what role work—defined for now as paid employment—plays, or should play, in our lives. This topic is more thoroughly dealt with in Nan Craig's contribution to this volume, but I will cover it briefly here since the WTR discussion intersects closely with the wider debate about whether the redundancy of human labour due to automation is desirable. This debate can be crudely split into two camps. The first camp argues that humans need work in order to lead meaningful lives. Automation can therefore be seen as a threat, if it is to lead to a reduction in the amount of work available. The second camp perceives work to be a burden, arguing that we would all lead more fulfilling lives without it. At the most radical end of this school of thought is accelerationism, a utopian leftist theory that the pace of automation should be sped up as much as possible in order to achieve a society in which human labour is unnecessary.[2]

I would argue, perhaps predictably, that the answer is more nuanced than either camp would have it. Those in the first camp often draw examples from areas in the UK affected by deindustrialisation in which whole communities have lost their jobs and incomes, and with these their sense

---

[2] Recent commentators to advocate this include Nick Srnicek and Alex Williams (2015) and Aaron Bastani (2018).

of purpose and self-respect. However, this outcome is specific to the circumstance: it does not mean that work is inherently necessary for meaning. The sudden removal of paid employment will inevitably lead to a loss of meaning for those affected, since there are no alternative structures of meaning available to supplant it. I suggest that it is not paid employment *per se* that gives our lives meaning. Rather, I draw upon Miroslav Volf's notion (1991) that work is 'instrumental activity'—activity towards a certain end—whether paid or not. Any activity that places us into an ongoing and purposeful relationship with the world around us—whether socially, materially, spiritually or in any other way—will necessarily be meaningful. The gradual reduction of time spent in paid employment, rather than the sudden and total abolition of paid employment (which is highly improbable in any case) will allow such structures of meaning to be built up slowly, so that they accrue social legitimacy.

This sociological debate is nonetheless irrelevant in practice unless two things occur: new automation technologies must be adopted, and if adopted, these technologies must result in reduced demand for human work. Neither of these two things are a given. Firstly, in recent years the UK has adopted new automation technologies significantly less than some other European countries (e.g. Denmark, Germany). The take-up of automation technologies is dependent upon a number of factors: it doesn't simply happen because the technology exists. For one, the substitution of capital for labour is unlikely to happen unless it is profitable. The UK's recent slow growth in productivity has been attributed partly to its slow take-up of these technologies (Bailey and Harrop 2018).

Regarding the second requirement, the productivity gains from new technologies do not automatically accrue to workers in the form of WTR or even higher wages. As I noted earlier, increases in productivity over the last three or four decades have not led to WTR in the UK. If a connection is to be forged between automation and increased leisure time, targeted policies are needed. Furthermore, it is by no means the case that new technologies will lead to less demand for work *overall*: past cases of how technological change has created new employment opportunities are well-documented (see e.g. Frey 2019). If workers are displaced, the most important consideration is whether they are able to transition to new types of jobs without difficulty. As Keune and Dekker observe, the impacts of automation are complex and vary significantly across sectors (2018).

It is clear, then, that WTR will not happen by default as a result of new automation technologies. Changes in policy are needed if we think WTR is desirable—and I have argued that it is. If the UK is to reduce working time, how should this be carried out? This is where it becomes useful to ask how and why other industrialised countries have come to work fewer hours than the UK as well as to examine past attempts to reduce working time.

Case Studies

There are several methods through which shorter working has been achieved in the past: legislation, collective bargaining, company-level initiatives, and individual and voluntary time reduction. I take the Netherlands, Germany and France as case studies, each of which uses different combinations of these approaches and with varying success.

Full-time employees in the Netherlands work 39 hours on average, less than the average full-time British worker. This is mainly due to shortening working hours through collective bargaining. More notably, however, an unusually large number of employees work voluntarily part-time (as opposed to involuntarily working part-time when they would rather work more hours): throughout the 1980s and 1990s there was a huge rise in part-time work (De Spiegelaere and Piasna 2017). Taking full-time and part-time employees together, the Dutch work a weekly average of under 30 hours (ibid.). The Netherlands has the highest percentage of part-time workers in Europe. More than 70 per cent of all working women and over one in four men have a part-time job. This is a rare (if not unique) example of individual, voluntary time reduction on a mass scale.

Part-time working in the Netherlands has been encouraged by legislation that has given part-time workers equivalent rights to full-time workers. For example, part-time workers are now entitled to the same minimum holiday allowance and training as full-time workers. Furthermore, under the Working Hours Adjustment Act of 2000, Dutch full-time workers have the right to reduce their work hours, while part-timers who want more work can adjust their hours upwards. Employers can only refuse if they can show that significant business or organisational interests stand in the way. However, voluntary time reduction only works when employees are sufficiently well-paid to have a preference for increased leisure over higher wages. This would not be feasible for many UK workers, around 18 per cent of whom are currently classified as low-paid. Through legislation around workers' rights, then, the Netherlands has been able to change public norms relating to part-time work by reducing the bias against it.

By contrast, Germany has mainly achieved WTR through collective agreements at the industry level. These have continued up to the present day. As of 2004, sector-level collective bargaining had resulted in a 35-hour week for a fifth of German workers (Hayden 2013: 129). Recently, workers represented by the German railway and transport union Eisenbahn- und Verkehrsgewerkschaft (EVG) have gained the opportunity to choose between wage increases and more holidays. In 2017, 56 per cent of employees at the rail company voted in favour of boosting their holiday allowance by six days. This was, again, only possible because employees were already satisfied with their wages. Germany has much stronger collective bargaining institutions than the UK, where during the 1980s most employer federations were dismantled or ended their involvement in collective bargaining. The UK now has predominantly company-level bargaining: there is still industry-level bargaining in some industries, such as parts of the textile industry and construction, but in most cases, in the private sector, bargaining is at company or plant level. This is less effective than sectoral bargaining as agreements cover many fewer people. Furthermore, the UK has low union density (i.e. union membership) in the private sector, at around 13 per cent.

France has been the only country in recent years to legislate at a nationwide level for a shorter working week. It is difficult to assess the long-term effects of this legislation—which reduced the working week to 35 hours—as it in fact consisted of two bills (1998 and 2001), which were then weakened by subsequent conservative governments introducing counter-reforms.

The 1998 bill set out that a reduction in hours was to operate without a reduction in salaries. The first WTR law set new legal working hours at 35, to be reached by the year 2000 by companies with more than 20 employees. It made provision for financial support for companies that negotiated the WTR agreements and maintained or increased employment levels. The primary objective was the creation of jobs. However, employers opposed this bill. They signed voluntary agreements to reduce working time in exchange for greater flexibility in working hours and a low level of job creation. The second bill of 2001 arose from the opposition of employers.

The legislation had two objectives: primarily to reduce unemployment, and secondly to improve quality of life. These objectives were partially achieved, but not to the extent hoped; 350,000 jobs were created but this was less than was initially forecasted. One reason was that many companies simply intensified work. Increased work-time flexibility—for example working evenings and weekends—reduced quality of life for some (Hayden 2013: 128). On the other hand, a survey conducted in 2001 found that parents with children under the age of 12 were overwhelmingly in favour of WTR (Méda 2013).

It is apparent from these case studies that institutions and norms play a central role in determining working hours. Cultural norms are not fixed and can be altered through institutional change, as the Netherlands case shows, since part-time work has not always been so widely accepted there. These case studies also show that when it comes to the success of WTR, the 'devil is in the detail', as De Spiegelaere and Piasna observe (2017: 66). We therefore need to examine the specifics of the UK labour market closely in order to gauge which policies might be appropriate.

Challenges for the UK Labour Market

At present, there are a number of problems in the UK labour market that, I would suggest, need to be addressed before legislating for WTR is considered. As is much discussed at present, the UK's productivity growth is low and real wages have been stagnant since 2008. Whilst the proportion of low-paid workers[3] has fallen since the introduction of the National

---

[3] Defined as those earning less than two-thirds of the median wage.

Living wage in 2016, it is still relatively high at 17.8 per cent according to the latest Office for National Statistics publication. Needless to say, shortening statutory working hours will not help those who are low-paid and/or underemployed.

Atypical and 'precarious' work has become more prevalent in recent years, although, as David Coats rightly notes, the so-called Uberisation of the economy has not occurred to the extent that media headlines often suggest (2018: 76). People in full-time work with permanent contracts still make up the majority of the labour force: the percentage of such workers dropped from 65 per cent to 63 per cent between 2008 and 2010 and has remained constant since. Nonetheless, it is not clear whether or how a statutory shortening of working time would affect gig-economy workers.

As mentioned, union membership in the UK is low, and there is a near total absence of sectoral bargaining: this makes it harder to translate the gains from automation into higher wages and/or more leisure time. Furthermore, the UK is already suffering from a shortage of technical skills: employers are struggling to fill hundreds of thousands of positions. The manufacturing and construction sectors are particularly affected by this shortage (Aubrey 2018: 240). The UK cannot take advantage of the benefits of automation unless workers have the skills to make it possible.

Finally, an inevitable challenge for a WTR agenda in any country is that the effects of WTR vary depending on the occupation in question. In occupations where presence, rather than productivity, is required (e.g. security guards, hospitality workers), reducing hours will not be compensated for by productivity gains: it will necessitate hiring more people, which has to be financed somehow. In others, such as social work, WTR may result in prioritising efficiency at the cost of the quality of service. It may be that the pursuit of WTR is not possible or desirable in all occupations.

Policy Recommendations

As this essay has shown, the question of how we organise our working time intersects with a wide range of the problems society faces today. The purpose of this essay, however, is not to provide a comprehensive policy roadmap, but rather to further the conversation around working time. I will therefore instead make one policy suggestion as an example of how institutions should be a central focus for a WTR agenda.

My suggestion is that the UK needs to set up new social partnership bodies in order to manage the connection between automation, wages and working time. As discussed, low pay is a persistent problem in the UK. The interests of low-paid workers were formerly represented on the Wages Councils, which were composed of an equal number of employers' and workers' representatives, and would set wage rates in weakly unionised sectors. However, these were abolished by John Major's government in 1993. I suggest the establishment of institutions that are similar in concept to the Wages Councils but with modifications in order that they be better suited to addressing current challenges, such as automation. Like the Wages Councils, these bodies would be sector-specific. They would bring together employers, workers' representatives, trade unions and government in order to develop strategies to increase productivity through targeted investment in the sector in question and tie those productivity gains to reduced working hours without cutting pay. In low-pay sectors, increasing pay rather than reducing working hours might be the initial priority.

These bodies would need to have legitimacy with both employers and workers. Coats (2018) and Brown and Wright (2018) both make the point that the support of employers is essential for this type of institution to function. For this reason, a resurrection of sectoral collective bargaining, much discussed at present, will not be successful if it is simply imposed upon employers, since employers are now accustomed to setting wages and working conditions independently (ibid.: 3). I would suggest that the inclusion of productivity growth in the agenda for these social partnership bodies would help in this regard, since this is in the interest of employers.

This is, of course, only one suggestion in isolation. Due to the interconnected nature of working time, many more elements need to be addressed in order to put together a convincing policy proposal for reducing working time. My point of departure is to acknowledge the central role of institutions in determining working norms, and the importance therefore of creating legitimate and effective institutions that will produce and perpetuate the norms we wish to see.

References

Aubrey, T. (2018). Why Adaptive Technical Skills Systems are Needed to Capitalise on the Technological Revolution: Challenges for the UK. In M. Neufeind, J. O'Reilly, & F. Ranft (Eds.), *Work in the Digital Age*. London: Policy Network.

Bailey, O., & Harrop, A. (2018). UK: Preparing for a Digital Revolution. In M. Neufeind, J. O'Reilly, & F. Ranft (Eds.), *Work in the Digital Age*. London: Policy Network.

Bastani, A. (2018). *Fully Automated Luxury Communism*. London: Verso Books.

Bosch, G. (2000). Working Time Reductions, Employment Consequences and Lessons from Europe: Defusing a Quasi-religious Controversy. In L. Golden & D. Figart (Eds.), *Working Time: International Trends, Theory and Policy Perspectives*. London: Routledge.

Brown, W., & Wright, C. (2018). Policies for Decent Labour Standards in Britain. *Political Quarterly, 89*(3), 482–489.

Coats, D. (2018). *Fragments in the Ruins: The Renewal of Social Democracy*. London: Policy Network.

D'Arcy, C. (2018). Low Pay Britain 2018. *Resolution Foundation*. Retrieved from https://www.resolutionfoundation.org/app/uploads/2018/05/Low-Pay-Britain-2018.pdf

De Spiegelaere, S., & Piasna, A. (2017). *The Why and How of Working Time Reduction*. Brussels: ETUI.

Frey, C. (2019). *The Technology Trap*. Princeton University Press.

Hayden, A. (2013). Patterns and Purpose of Work-Time Reduction—A Cross-national Comparison. In A. Coote & J. Franklin (Eds.), *Time on Our Side: Why We All Need a Shorter Working Week*. New Economics Foundation.

ILO. (2019). Global Commission on the Future of Work.

Keune, M., & Dekker, F. (2018). The Netherlands: The Sectoral Impact of Digitalisation on Employment and Job Quality. In M. Neufeind, J. O'Reilly, & F. Ranft (Eds.), *Work in the Digital Age*. London: Policy Network.

Méda, D. (2013). The French Experience. In A. Coote & J. Franklin (Eds.), *Time on Our Side: Why We All Need a Shorter Working Week*. New Economics Foundation.

Skidelsky, R., & Skidelsky, E. (2013). *How Much is Enough? The Love of Money, and the Case for the Good Life*. London: Penguin.

Smith, R. (2018). This Country Works the Longest Hours in Europe. *Weforum.org*. Retrieved from https://www.weforum.org/agenda/2018/02/greeks-work-longest-hours-in-europe/

Srnicek, N., & Williams, A. (2015). *Inventing the Future: Postcapitalism and a World Without Work*. London: Verso Books.

Stirling, A., & Lawrence, M. (2018). Time Banking: Bank Holidays, the Four-Day Week and the Politics of Time. *IPPR*. Retrieved from https://www.ippr.org/blog/time-banking-bank-holidays-the-four-day-week-and-the-politics-of-time

The Economist. (2013). Get a Life. Retrieved from https://www.economist.com/free-exchange/2013/09/24/get-a-life

Van Wanrooy, B., et al. (2014). The 2011 Workplace Employment Relations Study. *Gov.uk*. Retrieved from https://assets.publishing.service.gov.uk/government/uploads/system/uploads/attachment_data/file/336651/bis-14-1008-WERS-first-findings-report-fourth-edition-july-2014.pdf

Volf, M. (1991). *Work in the Spirit: Toward a Theology of Work*. New York and Oxford: Oxford University Press.

18

Shaping the Work of the Future: Policy Implications

Irmgard Nübler

An important lesson to be drawn from history is that the impact of new technologies on the future world of work is non-deterministic, it needs to be shaped, and societies and governments have choices. Moreover, experience shows that countries differ in innovation behaviour, and thus also in the impact of new technologies on labour markets and employment. While market forces play an important role, they are embedded in societies. This highlights the fundamental role of institutions, socially shared knowledge and belief systems, attitudes and aspirations of society in shaping the future of work.

This paper will first provide a broad framework for the analysis of policy implications. It views technological change as a complex, uncertain, costly and non-linear process, and explains the forces destroying, creating and transforming jobs. Secondly, a wide range of policies will be discussed that promote learning, innovation and economic transformation as well as adjustment in labour markets and thus the dynamics of jobs creation in the context of technological change.

I. Nübler (✉)
International Labour Organisation, Geneva, Switzerland
e-mail: nubler@ilo.org

The Framework

The framework takes into account three distinct, but related concepts to explain how new technologies link to the work of the future, thus raising a wide set of policy issues.

Process Innovation, Product Innovation and Market Forces

This paper takes a broad definition of technological change and innovation in order to demonstrate the impact of different types of innovation on the future world of work. While technology is defined as useful knowledge, innovations relate to new ideas of entrepreneurs for commercial use, and their implementation in the economy. Moreover, following Schumpeter (1911), we distinguish between process innovations and product innovations. Process innovations relate to new ways of producing goods and services, and new organization of work or business models. In contrast, product innovation is expressed in product differentiation, the implementation of significantly improved quality and the development of fundamentally new products, industries and sectors.

Historical experience shows that market forces play an important role in driving process innovations. In competitive markets, entrepreneurs are under pressure to increase productivity, which they achieve by introducing labour saving technologies. The quest for productivity is in particular high in the industrial production mode where firms compete mainly in price, costs and quality. As a result, since the Industrial Revolution, automation and fragmentation of production systems are long-term trends of technological change. They enhance productivity and competitiveness by saving labour and thereby destroying jobs. Process innovations are associated with declining employment, however, increasing complexity of occupations and jobs profile.

Market forces also create jobs, and can at least partially compensate for job losses. Various adjustment mechanisms are identified. Technology-induced productivity growth, if translated into higher wages, and into reduced prices, will enhance demand for domestic products and expand output. Furthermore, higher profit stimulates investment that will lead to further productivity gains through innovation and scale economies.

Moreover, as higher productivity is translated into increasing wages and declining working hours, demand for leisure-related activities increases, which since the industrial revolution has led to the development of entire new leisure industries and services, and thereby to the creation of new jobs. Since the leisure industries also adopt new technologies, the new jobs tend to become more sophisticated and skills intensive. The critical assumption in this argument is that the productivity gains arising from process technologies are shared with workers and consumers, and thus increase demand and local production (Vivarelli 2014).

Another major channel of job creation is the rise of new capital, software and robotics industries. The same process innovations that displace workers in the user industries create demand for workers in the producer industries. The new robots and smart machines need to be developed, designed, built, maintained and repaired. Additionally, the Internet of Things, Industry 4.0, digital Taylorism, driverless cars, big data and artificial intelligence require high investment in new infrastructure such as broadband, transport equipment and IT equipment, as well as increasingly complex software.

As a result, process innovations and compensation effects destroy and create jobs, however, they tend to create fewer jobs than they destroyed. Also, the new job profiles and occupations emerging with product innovations tend to be more complex and skills intensive. Capital and human capital become complementary which is also reflected in the current rise in demand for scientists, engineers and technicians, and the development of new occupations in particular at the intersection of professions, software and machines—Big Data architects and analysts, cloud services specialists and digital marketing professionals.

Shifting Techno-Economic Paradigm: Societal Demand and Political Choices

History shows that technological change is a dynamic, non-linear and long-term process. Technological change has been compared to waves which follow a particular dynamics. The current wave is defined by the digital technologies that were triggered by the invention of the microprocessor in the early 1970s, and is reflected in the new Information and Communication Technologies (ICTs) such as the internet, as well as robotics, Big Data and Artificial Intelligence (AI). The previous wave started in the early twentieth century with the diffusion of electricity grids and combustion engines.

Dosi (1982), Freeman and Perez (1988) and Perez (2002) describe the dynamics of technological waves as shifting technological or techno-economic paradigms. Following Perez, each paradigm is described by three distinct phases, and each of these phases is marked by distinct recurrences which allow one to identify the beginning and end of a phase. During the first phase, process innovations dominate. Enterprises and workers are learning the new and unlearning the old ways of doing things. Mechanization, automation and most recently robotization increase productivity, however, this phase is also characterized by unintended consequences: technological unemployment and anxiety, high and rising inequality (unequal distribution of productivity gains between labour and the owners of capital and skills), financial bubbles and crises as well as intensifying tensions between the existing institutions and the requirements of the economy and labour markets.

The second phase which may also be referred to as the "Golden Age" tends to be characterised by product innovations. Entrepreneurs search for new activities to create value, generate high innovation and imitation dynamics, and new growth industries replace existing ones. New institutions support these transformative changes and diversification of production structures, while societies develop a new consensus on the future lifestyles and consumption patterns. Technological change and innovations are perceived to be positive because they create new and better jobs and occupations. Finally, a third phase is characterized by low innovation activities, and low productivity growth.

The most critical period in sustaining the dynamics within a technological paradigm is the transition between the first and second phases. This transition is not automatic, and it cannot be generated by markets. The reason is that such fundamental changes in the economy can only be driven by new social demand and new political choices. Polanyi

(1944) argues that the development of a new economic system has always been accompanied by a change in the organization of society itself. For example, the evolution of market economies was underpinned by the emergence of market societies.

The important question then is, what triggers social transformation and new social and political demand? History shows that unintended consequences of the first phase of a technological paradigm (unemployment, inequality, financial and economic crisis) has generated in societies feelings of anxiety, injustice and uncertainty, as well as concerns about political instabilities, and loss of trust in existing institutions. These disruptive effects in society mobilize counter-movements. Polanyi argues that while the self-regulating markets brought boundless and unregulated changes to societies, human society would have been annihilated if it had not created protective counter-moves. These resulted in new institutions such as trade unions and factory laws (Polanyi 1944).

Collective Capabilities to Innovate

Finally, the concept of social capabilities is discussed to explain the important role of societies in driving diversification and the transition into the Golden Age, and the empirical observation that countries differ significantly in their patterns and pace of innovations, and thus, in their net-jobs creation. Mainstream economic literature highlights differences in factor endowment, productive capacity, industrial structure and comparative advantages in explaining differences across countries as these factors determine cost structures and, therefore, which technologies and products are profitable. However, the more fundamental issue is what enables a society to create those institutions that allow labour markets to adjust to disruptive impact of process innovations, to mobilize creativity and entrepreneurial spirit, develop new products, industries and jobs, and to effectively trigger and manage economic transition.

Social capabilities are discussed as important drivers of dynamic development processes, and of job-creating innovation processes (Abramovitz 1986; Lall 1992; List 1841). The International Labour Organization (ILO) has developed a knowledge-based theory to explain where capabilities reside, how they are created, and how they shape structural and technological change (Nübler 2014). Based on theories of knowledge developed in different disciplines, it argues that innovation capabilities reside in the knowledge base of societies. They exist at the collective level of societies, not in the knowledge or skills of individuals. The capabilities of societies to develop new products are embedded in the mix of knowledge and skills. Following evolutionary approaches to diversification, we describe a product as a combination of different sets of knowledge and skills required for its production, and therefore, the particular mix of skills and knowledge in a society determines those technologies, products and industries that a country can easily develop. As a general principle, the more diverse the technical knowledge sets and competences in the labour force, the wider the range of feasible products which enterprises may develop. For example, innovations need team players, communication skills and a diverse set of technical skills, but also workers who have the ability to focus on details, are persistent in searching for solutions and like to work individually.

While the mix of knowledge determines the feasible patterns of product innovation and structural transformation, the rules, procedures and collective know-how embodied in institutions determines a society's capability to manage processes of change, search for new solutions, mobilize creativity, entrepreneurship, and craftsmanship, and guide the choices and behavior of people. For example, Acemoglu and Robinson (2012) have shown that inclusive institutions benefiting large parts of society have also been central drivers of technological change. Schumpeter (1911) highlighted the important role of social institutions in explaining differences across societies in "entrepreneurial spirit" and their ability to drive processes of creative destruction. Moreover, "smart" apprenticeship institutions enforce high-quality training in broad competences of an occupation, craftsmanship and high status of craftspeople in society (Nübler 2014).

Socially shared belief systems such as cultures, ideologies, religion or philosophies determine the nature of such institutions and therfore, changing institutions for higher innovation capabilities also requires a change in these belief systems. For example, during the 1930s/1940s, a so-called "consumer society" evolved in the United States leading to new institutions such as consumer credits and commercials, changing consumption patterns and lifestyles, and massive demand for consumer goods, thus creating product innovations, new industries and jobs. This change was driven by a new belief system that provided great promise of unlimited happiness, freedom and social status in return for consuming goods and services. This belief system replaced one which valued thriftiness, and "being" rather than "having".

The framework presented in this chapter explains how technological change influences the work of the future. Technological change is explained as the result of a deliberate search process to solve problems and respond to economic, social and political demand, and it distinguishes between process innovation and product innovations to explain the main mechanism behind structural economic change, job creation and declining unemployment (Dosi 1982). The framework explains two distinct job-creating dynamics which challenge policies today. First, jobs are created when labour markets adjust to the job-destroying effect of productivity-enhancing process innovations. Jobs are expected to be created in the technology, capital goods, research and development (R&D) and leisure-related activities and sectors. Second, jobs are created as a response to new social and political demand which are induced by changes in societal thinking, mind sets, aspirations and development goals. The performance of countries in transformation and job creation, however, depends on the capabilities of societies to innovate and manage the transition process (Nübler 2016).

Using this framework to analyse the current technological, economic and social trends, we conclude that many technologically advanced countries are at the turning point within the digital techno-economic paradigm. This challenges governments to mobilize creativity, capabilities and a new social consensus on the way forward as well as target particular research, development and innovations to shape patterns of productive transformation that meet new social demand.

Policies to Shape Work in the Future

The broad nature of the framework elaborated in this chapter raises a wide range of policy issues in relevant fields: learning of individuals and societies; guiding science, technology and innovation for job creation; and enhancing labour market dynamics.

A Comprehensive Learning Strategy to Match Skills Demand and Build Capabilities

Education and training policies are discussed as the most important response to technological change. The challenge is to impart in students the skills needed in the labour market of the twenty-first century (Karstgen and West 2015). The future skills needs, however, is highly uncertain, as we lack information on the nature and speed of technological change, and innovation behaviour in a particular country context. Moreover, countries differ significantly in their education and training systems and these differences are maintained by differences in culture and the value societies give to different forms and levels of education. For example, countries like Austria, Switzerland and Germany show a relative low share in the labour force of post-secondary education graduates. The reason is that many young people enter apprenticeship training which is valued by society and enterprises. In contrast, in countries where apprenticeship training receives low status, technical education largely takes place in schools which is reflected in high shares in the labour force of post-secondary education graduates. This is mainly the case in Anglo-Saxon and Latin-speaking countries (Nübler 2018). In other words, socially shared attitudes and mindsets (belief systems) attach different values to the various forms of learning. This implies that countries need to develop country-specific strategies to cope with technological uncertainties. For example, while some countries focus on forecasting technological change and the specific skills needed in the future, they tend to provide a system of life-long learning that provides skills to workers as new demand emerges due to technological change. In contrast, other countries train young workers and apprentices in a wide set of technical and core competences relevant in the digital technological paradigm which builds-in flexibility in workers and often allows to absorb innovations without much further training.

While education and training systems need to respond to skills demand in current and future labour markets, societies also need to generate a learning process that transforms the society's knowledge base and develops strong innovation capabilities. Such learning occurs in different places: in schools, universities, training centres, enterprises, production systems, or social networks such as families and communities, subsequently creating two distinct capabilities. On one hand, policies enrich the societal knowledge base by imparting more complex skills, and by enhancing the diversity of the skills sets. which widens opportunities of enterprises for diversification. Capabilities to adopt newly emerging artificial intelligence tools, smart production systems and a technology-related industry requires a wide set of different competences, e.g. understanding and using Big Data, and the technical skills to perform accurate data mining and analysis. It includes enhanced Science, Technology, Engineering and Mathematics (STEM) education, but also liberal arts education and ethics, and learning to tolerate ambiguity when computers make decisions. For example, as AI-based systems are used across healthcare, criminal justice, finance and media, societies will face increased ethical and governance challenges.

On the other hand, policies support the change of mind sets, social norms and institutions in a society; generate capabilities to manage structural change processes; accelerate diversification; and drive a dynamic innovation process. New institutions are said to be "smart" when they influence education and occupational choices of students, consumption behaviour of consumers and innovation behaviour of entrepreneurs in a way that creates many good jobs in new activities, while also protecting the natural environment. For example the adverse consequences of environmental degradation have generated concerns within many societies about sustainable consumption behaviour and capitalistic production systems. The debate challenges existing development models, and the United Nations' Sustainable Development Agenda is an important step towards a new consensus (UNDP 2015). Moreover, while technological change has always had disruptive effects, societies were able to create a new consensus, forge a new societal contract and develop a new vision on the way forward. Such a consensus is based on a strong sense of justice, and trust in institutions. In the light of a growing perception of social injustice currently observed in many countries, a new consensus and social contract need to be forged for new political choices and institutions that support transition into a new economy and new jobs. In this context, social dialogue plays an important role. This "meta" institution is at the heart of societal learning as it guides, manages, accelerates and sustains the complex process of collective learning and institutional change. It is through the process of trustful and constructive conversations that all partners gain a deep understanding of the challenges, limits and the possible ways forward, and can work towards a consensus on the future we want.

Targeting Science, Technology and Innovation for balanced progress in economic, social and environmental goals

Given the high pressure in global markets to automate production processes, and the rise of artificial intelligence and smart production systems, manufacturing and related services will only generate a limited number of jobs. At the same time, local and global societies face many challenges at the social, environmental and economic levels. Governments need to design and implement policies to meet these challenges, and provide sustainable solutions which will balance progress in economic, social and environmental goals (UNDP 2015). Such mission oriented policies may aim at influencing the direction of technological change, and target particular economic sectors to influence patterns of diversification and structural change in the economy, or they may influence consumption behavior.

Many studies show the wide use of deliberate and proactive technology and industrial policies targeting selected industries to achieve growth, development and employment goals (Salazar-Xirinachs et al. 2014). For example, the so-called green economy aims to protect the environment while also creating good jobs, and leap-frogging into strategic technologies can create steep learning curves and new innovation capabilities that are required to sustain the innovation dynamics. Moreover, policies to promote small and medium-sized enterprises, the "new artisan economy" and the crafts sectors, have a high potential to create new jobs in the middle skills level (Katz 2014).

In addition, science, technology and innovation policies need to address the huge social, environmental and health challenges faced by the global society. Environmental degradation such as climate change and plastic in the ocean, the need of safe drinking water for a growing global population, and growing multi-drug resistance of bacteria will require technological solutions. A global research and innovation programme needs to be launched, financed and coordinated by international and multi-lateral institutions. Artificial intelligence tools need to be developed in order to analyse big data sets and combine national data sets for new insights and solutions.

In other words, the role of governments is not only to create an enabling environment, or to fix market failures. They play a key role in targeting specific research, technology and innovations in order to balance progress in economic development and employment, social inclusion and decent work as well as environmental integrity (ILO 2019).

Manage Labour Market Dynamics: Shared Gains and Just Transitions

A high dynamics of labour market adjustment towards a new equilibrium and job creation to compensate for the jobs loss is based on the assumption that the productivity gains of technological change is shared with workers, consumers and creative entrepreneurs. In reality, however, this is not always the case. The high and rising inequality we currently observe in many countries limits purchasing power and demand and thus local production and jobs creation. High concentration of productivity gains in the hands of a few constrains markets in the adjustment dynamics and job creation. Fiscal and wage policies are instrumental in sharing the benefits of technological change with workers, while competition policies help to lower prices for consumers. Both policies increase purchasing power and demand for existing and new products. This effect is strengthened by policies intended to distribute work more equally within the labour force, for example, by reducing working time. Redistribution of productivity gains to creative entrepreneurs supports investment in start-ups, which can contribute to the development of a new industry and jobs in the software and robots industry.

In addition, labour market policies play an important role in managing a just transition of the workforce, and ensure a fair distribution of burdens associated with technological change. Policies need to mitigate the adjustment costs, and accelerate the transition of workers into new jobs. Active labour market policies support workers in enhancing their employability in future labour markets, social protection smoothens consumption and allows workers to search for new productive employment, while also investing in skills and employability. In this context, the "universal basic income" is discussed as a counter-policy measure. Moreover, fiscal policies may help slowing down job displacement and provide space and time to learn and adjust. Taxing robots is discussed as one possible instrument. In addition, institutions are needed that regulate decent work and generate fairness where artificial intelligence-based technologies emerge. Existing frameworks for policy, law and human/workers' rights may not adequately address these challenges, and negate artificial intelligence-related risks that may encourage the deepening of existing and widening inequalities and biases in recruiting and promoting staff.

Conclusion

A fundamental message is that markets alone cannot achieve the economic and societal transformation that will shape the new jobs and the nature of work people will aspire in the future. It requires deliberate choices and policies, and those countries that proactively shape this process will create good jobs. Managing the process of technological, social and economic transformation to shape the future of jobs requires a comprehensive strategy which involves the forging of a new social consensus on the way forward, transforming the social knowledge base and capabilities, and investing in activities and new growth sectors to meet economic, social and environmental development goals. While many studies in the recent past focused on estimating expected jobs losses due to technological change, the more relevant issue is how to design and implement policies that can effectively transform societies and economies for the creation of new and good jobs, and decent work while maintaining environmental integrity.

References

Abramovitz, M. (1986). Catching Up, Forging Ahead, and Falling Behind. *The Journal of Economic History, 46*(2), 385–406.
Acemoglu, D., & Robinson, J. S. (2012). *Why Nations Fail. The Origins of Power, Prosperity, and Poverty*. New York: Crown Publishers.
Dosi, G. (1982). Technological Paradigms and Technological Trajectories. "A Suggested Interpretation of the Determinants and Directions of Technical Change". *Research Policy, 11*(3), 147–162.
Freeman, C., & Perez, C. (1988). Structural Crises of Adjustment, Business Cycles and Investment Behaviour. In G. Dosi (Ed.), *Technical Change and Economic Theory* (pp. 38–66). London: Francis Pinter.

Karstgen, J., & West, D. (2015). New Skills Needed for New Manufacturing Technology. *Brookings.edu*. Retrieved August 30, 2016, from https://www.brookings.edu/blog/techtank/2015/07/15/new-skills-needed-for-new-manufacturing-technology/

Katz, L. (2014, July 15). Get a Liberal Arts B.A., Not a Business B.A., for the Coming Artisan Economy. *Pbs.org*. Retrieved August 30, 2016, from http://www.pbs.org/newshour/making-sense/get-a-liberal-arts-b-a-not-a-business-b-a-for-the-coming-artisan-economy/

Lall, S. (1992). Technological Capabilities and Industrialization. *World Development, 20*(2), 165–186.

List, F. (1841). *Das nationale System der politischen Oekonomie*. Stuttgart and Tübingen: J. G. Cotta'scher Verlag.

Nübler, I. (2014). A Theory of Capabilities for Productive Transformation: Learning to Catch Up. In J. M. Salazar-Xirinachs, I. Nübler, & R. Kozul-Wright (Eds.), *Transforming Economies: Making Industrial Policy Work for Growth, Jobs and Development* (pp. 113–149). Geneva: ILO.

Nübler, I. (2016). *New Technologies: A Job-Less Future or a Golden Age of Job Creation?* Working Paper, Research Department, International Labour Organization, No. 13.

Nübler, I. (2018). New Technologies, Innovation, and the Future of Jobs. In E. Paus (Ed.), *Confronting Dystopia—The New Technological Revolution and the Future of Work* (pp. 46–75). Ithaca and London: Cornell University Press.

Perez, C. (2002). *Technological Revolutions and Financial Capital: The Dynamics of Bubbles and Golden Ages*. London: Edward Elgar.

Polanyi, K. (1944). *The Great Transformation: The Political and Economic Origins of our Time*. New York: Farrar & Rinehart.

Salazar-Xirinachs, J. M., Nübler, I., & Kozul-Wright, R. (2014). *Transforming Economies: Making Industrial Policies Work for Growth, Jobs and Development*. Geneva: ILO.

Schumpeter, J. (1911). *The Theory of Economic Development: An Inquiry into Profits, Capital, Credit, Interest, and the Business Cycle*. Piscataway, NJ: Transaction Publishers.

UNDP (United Nations Development Program). (2015). Transforming Our World: The 2030 Agenda for Sustainable Development. Retrieved from https://sustainabledevelopment.un.org/post2015/transformingourworld

Vivarelli, M. (2014). Innovation, Employment and Skills in Advanced and Developing Countries: A Survey of Economic Literature. *Journal of Economic Issues, XLVIII*(1), 123–154.

Index[1]

A

Abramovitz, Moses, 193
Accelerative thrust, 159
Accountancy (automation of), 84, 114
Acemoglu, Daron, 194
Action, 2, 58, 74, 128
Agrarian societies/agrarian revolution, 27, 28
Agriculture, 3, 11, 39, 41, 44, 45, 47
Airbnb, 136, 137
Algorithms
    accountability, 140, 148
    accuracy, 148
    bias, 148
    ethics of, 6
    feedback loops, 146, 148
    gaming, 148
    prediction, 139, 146, 147
Alibaba, 136, 137
Alienation, 57, 58, 61
AlphaGo, 90, 112
Amazon
    Alexa, 140
    Web Services, 134, 140
Anthropomorphisation, 110
Apple, 73, 137, 138
Applebaum, Herbert, 74
Architecture, 43
Arendt, Hannah, 74
Arkwright, Richard, 29
Art, 61, 76, 107, 115–117, 119, 120, 170, 197

---

[1] Note: Page numbers followed by 'n' refer to notes.

Artificial Intelligence (AI)
   and art, 107, 115–117,
      119, 120
   and game-playing, 112, 115
   general *vs.* specialised, 130, 140
   in fiction, 113, 115, 146
   superintelligence, 112, 113
   Weak *vs.* Strong, 99
Artisans, 12, 29, 38, 74, 93, 94
Attitudes to work, 1, 4,
      53–62, 73, 75
Aubrey, 184
Austria, 68, 196
Authenticity, 116
Authority, 120, 165
Automation
   restrictions on, 95
   speed of, 21, 137
   task automation *vs* job
      automation, 92, 93, 110, 141
Autonomous cars, 114, 115, 118
Autor, David, 59, 126
Autor Levy Murnane (ALM)
      hypothesis, 126–128, 131

B

Bailey, Olivia, 180
Bairoch, Paul, 44, 46
Banking (automation of), 87, 147
Bargaining, 68, 70, 177, 181, 182,
      184, 185
Bastani, Aaron, 179n2
Beckert, Sven, 44
Berger, Thor, 95
Bessen, James, 4
Blumenbach, Wenzel, 41
Bosch, Gerhard, 179
Bostrom, Nick, 112, 113
Bourgeois household, 39

Brain
   and AI, 113
   analagous to computer, 100, 103,
      104, 115
Brown, William, 185
Bullshit jobs
   psychological effects, 162
Bureaucracy, 169

C

Capitalism, 12, 17, 28, 53, 57, 58,
      61, 75, 135, 159
Capper, Phillip, 127, 128
Care work, 3, 48, 75, 117, 178
Carlyle, Thomas, 28
Catholic, 74
Central Europe, 38, 40
Centralisation, 69, 175, 176
Chalmers, David, 103
Chatbots, 91
Chen, Chinchih, 95
Chess (and AI), 112
China, 95, 135
Christian (view of work), 74, 75,
      161, 166
Clark, A, 60
Class, 13–15, 17, 30, 39, 43, 46, 47,
      118, 159, 160, 162, 165, 172
Classical economics, 54, 55
Climate change, 30, 198
Cloud computing, 139, 140
Coase, Ronald, 70
Coats, David, 184, 185
Collective bargaining, 68, 181,
      182, 185
Communism, 13, 57, 58, 61
Competition, 12, 16–18, 39, 91, 94,
      112, 115, 119, 139, 140,
      152, 199

Computational Creativity, 109, 115, 120, 121
Computer aided design (CAD), 34, 35
Computer programming, 100, 116
Computer revolution, 90, 94, 95, 99
Computers, 20, 34, 84, 86, 90, 92–94, 99–107, 110, 111, 115, 116, 120, 131, 134, 146, 147, 151, 197
Consciousness
  of AI, 110–111
  the hard problem, 103
  of humans, 105
  objective *vs.* subjective, 102, 103
Consumerism/consumer society, 30, 74, 161, 194
Consumption, 3, 5, 12, 13, 16, 19, 38, 41, 56, 59, 61, 62, 66, 85, 88, 166, 176, 192, 194, 197, 199
Contested concepts, 120
Cooperatives, 40, 61, 69
Craftsmanship, 3, 11, 35, 36, 39, 194
Craig, Nan, 4, 179
Creative work, 3, 48, 74
Creativity, 3, 5, 57, 91, 105–107, 110, 120, 121, 193–195

D
D'Arcy, Conor, 177
Data, 2, 84, 92, 107, 129, 130, 137–140, 146, 149, 150, 153, 178, 191, 197, 198
Davies, W.H., 31
De Spiegelaere, Stan, 181, 183

Deep Blue, 91, 112, 129, 130
Dekker, Fabian, 180
Deliveroo, 136
Demand
  effects on automation, 4, 21, 86
  elasticity, 86
  of work, 4, 13, 15, 16, 76, 158, 164, 180, 199
Democracy, 28
Denmark, 68, 177, 180
Dennett, Daniel, 100, 102, 103
Developing countries, 145
Digital economy, 5, 19, 125–132, 140
Digital revolution, 70
Division of labour, 11, 35, 38, 43, 44, 55
Donkin, Richard, 3
Dosi, Giovanni, 192, 195
Do what you love, 73, 74, 76
Dreyfus, Herbert, 100

E
Economics, 1, 4, 5, 7, 10, 12, 14, 15, 18, 29, 30, 53–62
Economic view of work, 53–62
Education, 41, 42, 48, 67–69, 126, 131, 169, 171, 196, 197
Efficiency, 5, 16, 75, 159, 168, 184
Empathy, 106, 107
Employment
  law, 68
  rates, 67, 68, 70
English East India Company, 44
Entrepreneurs, 29, 70, 77, 190, 192, 197, 199
Environment, 25, 31, 56, 70, 87, 91, 109, 111, 113, 120, 178, 198

Equality
  of opportunity, 69
  of outcome, 69
  social, 163
Ethics
  of AI, 6, 110, 119, 145–153, 197
  stagnation of, 151–152
  of work, 28
Exit, 69
Experience, 36, 61, 85, 90, 94, 99–105, 116, 119, 189, 190

F

Facebook, 136–141, 161
Factory system, 29–30
Families, 3, 26, 29, 37–48, 75, 76, 138, 159, 162, 178, 196
Feminist (arguments about work), 79
Finance, 48, 87, 170, 197
Fire, harnessing/discovery of, 29
Firestone, Shulamith, 159
Firms, 16, 17, 68, 70, 85, 87, 133, 148, 149, 151, 152, 168, 169, 172, 190
Flexicurity, 68
Ford, Henry, 30
Ford, Martin, 2, 59, 106
France, 4, 6, 66–70, 177, 181, 182
Franklin, Benjamin, 28
Freeman, Chris, 192
French Revolution, 43
Frey, Carl Benedikt, 4, 180
Friedman, Milton, 171
Fuzzy matching, 148, 149

G

Galbraith, JK, 66
GDP, 19, 178

Gender, 38, 43, 44, 48, 151, 178
Gendered division of labour, 38, 43, 44
Germany, 6, 177, 180–182, 196
Gig economy, 27, 184
Globalisation, 20, 30, 90, 95
Google
  Google Cloud, 140
  Google Home, 140
  Google Maps, 35
  Google Translate, 106
Google DeepMind, 112, 119
Gorz, A., 59
Graeber, David, 6, 76, 157, 161, 168
Greek ideas of work, 74
Growth, 2, 6, 7, 12, 25, 27, 30, 31, 55, 69, 75, 85, 86, 88, 110, 126, 128, 130, 135, 169, 176, 180, 183, 185, 190, 192, 198, 200

H

Happiness, 5, 62, 195
Harrop, Andrew, 180
Hassabis, Demis, 119
Hayden, Anders, 182, 183
Healthcare, 3, 87, 94, 117, 165, 197
Heterodox economics, 54, 56, 62
Hierarchy, 46, 48, 55, 69, 170
High-skilled jobs, 128, 134
Homejoy, 135
Homo economicus, 56, 57
Homo laborans, 3
Homo ludens, 3
Household economy, 4, 38–40, 45, 47
Housewives, 42, 43, 46, 47
Housework, 39, 40, 42, 44, 47
Hunter-gatherers, 11, 26, 27, 30

Idleness, 54
India, 44–47
Industrial Revolution, 2, 4, 14, 29, 37, 75, 93, 94, 175, 177, 190, 191
Inequality, 67–69, 86, 87, 192, 193, 199, 200
Informal economy, 47
Information technology, 86, 161
Infrastructure
 digital, 140
 physical, 103
Innovation, 6, 10, 14, 16, 18, 34, 67, 69, 189–199
 process innovation *vs.* product innovation, 16, 18, 190–191, 195
International Labour Organisation (ILO), 193
Internet of Things, 139, 191
Investment
 in capital, 114
 in skills, 70

Japan, 117
Jensen, C, 55
Job guarantee, 172
Jobs, Steve, 73
Journalism
 automation of, 118
 clickbait, 118
Juries, algorithmic selection of, 150, 153

Karstgen, Jack, 196
Kasparov, Garry, 91, 112, 129, 130
Katz, Lawrence, 198
Kennedy, John F., 160
Keune, Maarten, 180
Keynes, John Maynard, 6, 9, 11, 27, 60, 61, 160, 161, 176
King, Martin Luther, 171
Knowledge (tacit *vs.* explicit), 127
Komlosy, Andrea, 4, 75
Kubrick, Stanley, 26
Kurzweil, Raymond, 101, 103, 104
Kuznets, Simon, 190

Labour, 3, 10, 11, 13–16, 18–21, 29, 34–36, 38, 43–46, 55, 59, 65–70, 73–76, 85–87, 89, 90, 93, 94, 96, 114, 125, 126, 128, 130, 131, 141, 158, 165, 176–180, 183–184, 189, 190, 192–196, 199–200
Labour market polarisation, 67, 70, 126
Labour markets, 67, 68, 70, 87, 90, 96, 125, 126, 128, 130, 131, 141, 178, 183–184, 189, 192, 193, 195, 196, 199–200
Labour-saving effect, 86
Lall, Sanjaya, 193
Language translation, 105, 106
Latent Damage Act 1986, 127
Law
 automation of, 145, 152, 153
 ethics, 145–153
Lawrence, Mathew, 177
Layton, E., 58
Le Bon, Gustave, 101
Lee, Richard, 26
Legal search/legal discovery, 148–150

Leisure, 3, 10, 11, 19, 27, 48, 55, 56, 59–62, 65, 77, 79, 117, 118, 159, 161, 178, 180, 182, 184, 191, 195
Levy, Frank, 126
List, Friedrich, 193
Love, 55, 74, 76, 99, 103, 106, 112, 118
Low-income jobs, 96
Loyalty, 69
Luddites, 2, 14, 18, 35, 59, 94, 96
Lyft, 136

M

Machine learning, 59, 84, 90, 91, 96, 138, 139
Machines, 2, 5, 10, 12–15, 17, 19, 20, 35, 36, 38, 59, 84–87, 90–96, 99–103, 105–107, 109–121, 127–131, 138, 139, 145, 147, 148, 160, 168, 191
Machine vision, 120
Malthusian, 19
Man, Henrik de, 79
Management, 27, 30, 41, 69, 70
  management theory/ organisational theory (see also Scientific management)
Mann, Michael, 46
Manual work, 1
Manufacturing, 86, 87, 90, 94, 95, 176, 184, 198
Markets/market forces, 5, 6, 21, 38, 44–46, 67, 68, 70, 79, 85–88, 90, 96, 120, 125, 126, 128, 130, 131, 140, 141, 150, 152, 159, 164, 165, 171, 178, 183, 189–193, 195, 196, 198–200

Marx, Karl, 17, 18, 27, 56–59, 61, 62, 78
Matrimonial relationships, 37
McCormack, Win, 159
Meaning, 4, 9, 10, 19, 25, 54, 57, 58, 66, 73, 76, 78, 79, 84, 106, 116, 176, 180
Mechanisation, 15, 17, 19, 20, 192
Meckling, W., 55
Méda, Dominique, 183
Medical diagnosis (automation of), 128, 129
Menger, Pierre-Michel, 4
Mental labour, 3
Meritocracy, 28
Middle-income jobs, 90, 93, 94
Migration, 40, 47
Minimum wage, 67, 69
Mining, 26, 38, 197
Mokyr, J., 59
Monopolies, 6, 136, 138–140
Morals/morality, 48, 77, 159, 160, 162, 164, 166, 167
Moravec's paradox, 131
Murnane, Richard, 126

N

Nagel, Thomas, 100, 102
National Living wage, 184
Needs vs. Wants, 3, 30, 88
Neoclassical economics, 4, 55, 60, 62, 73
Netherlands, the/Holland, 6, 68, 151, 163, 177, 181–183
Network effects, 138
Networks, 45, 48, 138, 196
Neumann, John von, 99
New Zealand, 179
Nübler, Irmgard, 6, 194, 196

O

Obama, Barack, 164, 165, 171
Obligation, 38, 53, 73–79
Occupations, 16, 40, 41, 46, 47, 58, 70, 83, 84, 86, 87, 90, 92, 106, 178, 184, 190–192, 194
OECD, 66–68, 178
O'Neil, Cathy, 6
Ontology of work, 65
Organisations
 dynamics of, 164
Osborne, Michael, 90
Oswald, A, 60

P

Painting Fool, The, 115, 116, 120
Parenting, 75, 76
Patocka, Jan, 9, 21
Pattern recognition, 129
Peasant labour, 41
Perez, Carlota, 192
Philosophy of work, 30
Physical labour, 3
Piasna, Agnieszka, 181, 183
Piece-work, 30
Platform economy/platform capitalism, 6, 140
Polanyi, Karl, 192, 193
Polanyi, Michael, 127
Policy (argument against), 7, 21, 67, 68, 95, 157–173, 180, 181, 183–185, 189–200
Population, 2, 12, 15–17, 19, 28, 30, 89, 90, 117, 147, 158, 172, 198
Postmates, 136
Post-work society, 59
Poverty, 15, 47, 59, 67, 177
Pre-modern/pre-industrial work, 3, 11, 47, 48
Productivity, 7, 10, 79, 86, 87, 176, 178–180, 183–185, 190–192, 199
Professional work, 1, 39
Profits (different profit models), 14–18, 30, 48, 75, 79, 93, 134, 135, 138, 152, 191
Protestant work ethic, 28
Public services, 94, 167
Puritan (view of work), 28, 75, 166

R

Redistribution, 79, 169, 199
Redundancy, 10, 12, 15–17, 19, 78, 179
Religion/religious ritual, 12, 28, 194
Remittances, 40
Responsibility, 44, 47, 76–79, 106, 107, 115, 118, 136
Retail sector, 87, 137
Retirement, 19, 67, 78
Ricardo, David, 2, 13–17
Robinson, James, 194
Robotisation, 21, 94, 95, 192
Robots
 carers, 106
Romantic (view of work), 34, 35
Ruskin, John, 34

S

Safety nets, 67, 68
Sahlins, Marshall, 26, 158
Salazar-Xirinachs, Jose M., 198
Schumpeter, Joseph, 190, 194
Scientific management, 30
Scott, James C., 28

Searle, John, 100–103
Self-employment, 69–70, 75
Self-realisation, 57, 165
Sennett, Richard, 3
Services/service sector
  low frequency vs. high frequency, 134
  work, 40, 68, 161, 163
Singularity, 116
Skidelsky, Edward, 60, 176
Skidelsky, Robert, 60, 176
Skills
  acquisition, 33, 70
  skilled vs. unskilled labour/jobs, 67
Slavery, 11, 29, 30, 45
Smartphones, 140
Smiles, Samuel, 28
Smith, Adam, 12, 13, 27, 35, 54, 55, 65
Smith, Rob, 177
Social drawing rights, 70
Social interaction, 53, 88, 91
Social media, 77, 138, 168
Societal knowledge base, 196–197
Sociology (of work), 166
Spencer, David, 4, 54, 59, 61
Spinning mills (cotton industry?), 29
Srnicek, Nick, 5, 59, 179n2
Star Trek, 146–148
Status goods, 88
Stirling, Alfie, 177
Stoics (view of work), 74
Stradivarius, 33–35
Subsistence, 27, 38, 40, 41, 44, 45, 73, 75, 76
Summers, Larry, 2
Supply and demand, 16, 21
Susskind, Daniel, 5
Susskind, Richard, 127, 132

T
Tasks
  routine vs. non-routine, 126, 127, 129, 131
  simplification, 91, 92
Taylor, Frederick Winslow, 30
Technological determinism, 5
Technological progress, 9, 18, 59, 89, 93, 96, 131, 176
Technological unemployment, 2, 6, 10, 16, 160, 173, 192
Technology, 2–5, 7, 9, 16–19, 27–30, 35, 57, 59, 61, 62, 75, 83–96, 110, 111, 115, 117, 119, 120, 126, 129, 131, 133, 139, 140, 145, 149, 150, 160, 161, 180, 181, 189–195, 198–200
Terkel, Studs, 4
Textile industry, 85, 182
3D printing, 35
Time and motion studies, 30
Toffler, Alvin, 159
Tokumitsu, Miya, 73
Tools/tool-making, 11, 26–28, 34, 35, 70, 109, 149, 197, 198
Trades Union Congress (TUC), 175, 177
Trump, Donald, 94, 95
Turello, Dan, 103
Turing, Alan, 100, 105
Turing test, 91, 101n1

U
Uber, 6, 133–137
Uberisation (of the economy), 27, 133, 134, 184

Unemployment, 10, 11, 16, 17, 59, 60, 68, 78, 89, 160, 164, 171, 178, 179, 183, 193, 195
Unions, 68, 69, 136, 176–178, 182, 184, 185, 193
United Kingdom, 6, 26, 68, 127, 151, 163, 164, 175–185
Universal Basic Income (UBI), 70, 78, 171, 199
USA, 15, 28, 68, 83, 85, 86, 89, 126, 151, 165, 166, 178, 194
Utility, 55, 62, 94, 166

V

Value
  extraction of, 134
  labour theory of, 165
  of work, 11, 31, 58, 60, 61, 65, 66, 73, 163, 165–167
Van Wanrooy, Brigid, 178
Veblen, Thorstein, 27, 56–58, 62
Venture capital, 111, 114, 135
Violin making, 34
Vivarelli, Marco, 191
Vocational training, 68
Voice, 69, 106, 147, 159
Volf, Miroslav, 176, 180
Vonnegut, Kurt, 158, 160

W

Wages
  minimum, 69
  stagnation, 87, 89, 94, 183

Walsh, Toby, 119
Weaving industry, 18, 29, 38, 85
Weber, Max, 75
Weeks, Kathi, 79
Welfare, 5, 54, 60, 66–70, 135, 160, 171
Welfare state, 66, 69, 70, 160, 171
Welfarist understanding of work, 65
Wellbeing, 19, 27, 66, 177, 179
West, Darell, 196
Western Europe, 4, 37, 39, 44
Williams, Alex, 59, 179n2
Williamson, O, 55
Wilson, Frank, 34
Work
  as a cost/burden, 13, 18, 44, 55, 57, 58, 60, 75, 77, 78
  freedom from, 39, 60, 77, 78
  as meaningful, 76, 77, 179, 180
  as pleasurable, 3
Workforce skills, 6
Working hours
  increase vs falls in, 19
  part-time *vs.* full-time, 181
  targeted reduction of, 185
Working Hours Adjustment Act 2000, 181
'Working poor' model, 67, 68
Work-life balance, 79, 179
Wright, Chris F., 185

Z

Zuckerberg, Mark, 138

Druck:
Customized Business Services GmbH
im Auftrag der
KNV Zeitfracht GmbH
Ein Unternehmen der Zeitfracht - Gruppe
Ferdinand-Jühlke-Str. 7
99095 Erfurt

This short, accessible book seeks to explore the future of work through the views and opinions of a range of expertise, encompassing economic, historical, technological, ethical and anthropological aspects of the debate. The transition to an automated society brings with it new challenges and a consideration for what has happened in the past; the editors of this book carefully steer the reader through future possibilities and policy outcomes, all the while recognising that whilst such a shift to a robotised society will be a gradual process, it is one that requires significant thought and consideration.

Robert Skidelsky is Emeritus Professor of Political Economy at Warwick University, UK and Chair of the Centre for Global Studies, London, UK. His three-volume biography of John Maynard Keynes won five prizes and his book on the financial crisis, *Keynes: The Return of the Master*, was published in September 2010. He was made a member of the House of Lords in 1991, where he sits on the cross-benches, and was elected as a fellow of the British Academy in 1994. *How Much is Enough? The Love of Money and the Case for the Good Life*, co-written with Edward Skidelsky, was published in July 2012. His most recent publications were as author of *Britain Since 1900: A Success Story?* (2014), *The Essential Keynes* (2015), *Markets and Morals* (Palgrave Macmillan, 2015), *Who Runs the Economy* (Palgrave Macmillan, 2016), and *Austerity vs. Stimulus* (Palgrave Macmillan, 2017).

Nan Craig is Programme Director at the Centre for Global Studies, London, UK. She studied politics and international studies at the University of Warwick, UK, and global politics at the London School of Economics, UK. She co-authored *Who Runs the Economy?* (Palgrave Macmillan, 2016) with Robert Skidelsky.

ISBN 978-3-030-21133-2

www.palgrave.com